WATER: 2ND EDITION
A MATRIX OF LIFE

RSC Paperbacks

RSC Paperbacks are a series of inexpensive texts suitable for teachers and students and give a clear, readable introduction to selected topics in chemistry. They should also appeal to the general chemist. For further information on all available titles contact:

Sales and Customer Care Department, Royal Society of Chemistry,
Thomas Graham House, Science Park, Milton Road, Cambridge CB4 0WF, UK
Telephone: +44(0) 1223 432360; Fax: +44 (0) 1223 423429; E-mail: sales@rsc.org

Recent Titles Available

The Chemistry of Fragrances
compiled by David Pybus and Charles Sell
Polymers and the Environment
by Gerald Scott
Brewing
by Ian S. Hornsey
The Chemistry of Fireworks
by Michael S. Russell
Water (Second Edition): a Matrix of Life
by Felix Franks
The Science of Chocolate
by Stephen T. Beckett

RSC Paperbacks

WATER: 2ND EDITION
A MATRIX OF LIFE

FELIX FRANKS

Cambridge, UK

ISBN 0-85404-583-X

A catalogue record for this book is available from the British Library

Published by The Royal Society of Chemistry,
Thomas Graham House, Science Park, Milton Road,
Cambridge CB4 0WF, UK

For further information see our web site at www.rsc.org

Typeset in Great Britain by Vision Typesetting, Manchester
Printed by Athenaeum Press Ltd, Gateshead, Tyne & Wear, UK

Preface

Through the ages, the wondrous nature of water has inspired poets, painters, composers and philosophers. Water was one of Aristotle's four elements, and long after earth, fire and air had been recognised for what they were, water was still regarded as an element. It was not until the end of the 18th century that Lavoisier and Priestley demonstrated water to be a 'mixture' of elements. But some two hundred years before that, Leonardo da Vinci had already published his book 'Del moto e misura dell'acqua' in which he described sophisticated investigations into the physical properties of water. Most present day scientists, technologists and economists are not so easily impressed: they take water very much for granted and rarely spare a thought for its unique role in influencing the course of chemical reactions and for the shaping of our terrestrial environment to make it fit for life.

This book is an update on the RSC Paperback 'Water' which was published in 1983. I have tried once again to condense the present state of our scientific knowledge of water and aqueous systems, with emphasis on new insights, mainly to be found in the later chapters. The book also touches on some extra-scientific issues that are, however, of extreme importance to society: water quality, usage and management worldwide, its availability, economics and politics.

As before, the book is dedicated to my two mentors who shaped my career. David Ives taught me all about self-discipline in laboratory science and in written scientific communication, where every word should be carefully weighed before it is committed into print. Henry Frank opened my eyes to the universal wonders of water and taught me the methods of problem solving: 'You do not solve problems by hitting them over the head but by gradually backing them into a corner from which there is no escape'.

Although the proverb has it that a rolling stone gathers no moss, during my diversified career I have gathered a great deal of moss. It is a pleasure, therefore, to acknowledge my gratitude to the many colleagues who, directly or indirectly, have helped me to gain a better understanding of water with discussions, suggestions or practical collaboration. In addition to those mentioned in the 1983 Paperback, my very special thanks go to former colleagues Patrick Echlin, Tony Auffret, Barry Aldous, Toshiki Wakabayashi, Norio Murase, Kazuhito Kajiwara, Evgenyi Shalaev, Mike Pikal, Harry Levine and Louise Slade, for many years of productive collaboration and personal friendship.

Felix Franks
BioUpdate Foundation
Cambridge
April 2000

Contents

Acknowledgements

When I was invited by the Royal Society of Chemistry to write an updated version of the slim paperback 'Water', it occurred to me that, for the past 15 years I had been more involved with the removal of water (freezing and drying) than any other aspect. I therefore wrote to friends and colleagues, asking what had happened to water research in its broadest sense. They all very generously responded and pointed me in the right direction for further study. I am therefore greatly indebted to them and gratefully acknowledge the part they have played, although passively, in the production of this book.

Thank you, Randolph Barker, Pablo Debenedetti, John Finney, Donald Irish, Robert Wood, Jan Engberts, George Zografi, Colin Edge, Harry Levine, Louise Slade and Hiroshi Suga.

Chapter 1

Origin and Distribution of Water in the Ecosphere: Water and Prehistoric Life

The Eccentric Liquid

Water is the only inorganic liquid that occurs naturally on earth. It is also the only chemical compound that occurs naturally in all three physical states: solid, liquid and vapour. It existed on this planet long before any form of life evolved but, since life developed in water, the properties of the 'Universal Solvent' or 'Life's natural habitat' or 'Life's preferred habitat' came to exert a controlling influence over the many biochemical and physiological processes that are involved in the maintenance and perpetuation of living organisms. It is therefore in order to discuss briefly the occurrence of water on earth, its distribution and its controlling influence on the development of life.

The Hydrologic Cycle

At present we are left with several puzzles concerning the composition of the atmosphere and the quantity of water in the hydrosphere. It seems safe to assume that the presence of water in the *liquid* state can only date from a time when the temperature of the earth's crust had dropped to below the critical temperature of water, 374 °C. If all the water that now makes up the oceans had previously existed as a supercritical water atmosphere, then the pressure per square metre of earth surface would have been 25 MPa. During cooling to below the critical point, vast masses of water would have condensed onto the earth's surface and also penetrated deep into rock crevices. Some of this water would immediately have boiled off again, to be recondensed at a later time. The hydrologic evaporation–condensation cycle could thus have begun several billion years ago. It is therefore irrelevant whether the heat of the earth itself or solar radiation initiated the cycle. What is relevant, however, is the scale of the water movement. The total water content of the atmosphere is 6×10^8 ha m (1 ha m = 10 000 m^3). This is the amount of water which will cover an area of 1 ha to a depth of 1 m. Since the total annual precipitation is 225×10^8 ha m, the water in the atmosphere is turned over 37 times every year. This level of precipitation is equivalent to a water depth of 0.5 m averaged over the earth's surface. Such an averaging is of course meaningless, because the level of precipitation is quite non-uniform in time and space.

Although the hydrologic cycle, shown in Figure 1.1, is a continuum, its description usually begins with the oceans which cover 71% of the earth's surface. The heat of the sun causes water to evaporate. Under the influence of certain changes in temperature and/or pressure, the moisture condenses and returns to earth in the form of rain, hail, sleet or snow – collectively referred to as water of meteoric origin. It falls irregularly with respect to geographical location, with coastal areas receiving more.

Of the average rainfall, about 70% evaporates; the remainder appears as liquid water on or below the land surface. Some water evaporates in the air between the clouds and the land surface. The remaining losses are of two forms: direct evaporation from wet surfaces and transpiration through plants from their leaves and stems. The 30% of water not directly returned to the atmosphere constitutes the runoff and provides our potentially available freshwater supply. Actually, the proportion of the earth's total freshwater resources which participates in the hydrologic cycle does not exceed 0.003%; the remainder is locked up in the Antarctic ice cap. If melted, it would supply all the earth's rivers for 850 years.

Enormously large quantities of water participate in the cycle, as shown in Table 1.1. The oceans constitute by far the largest proportion of our water resources, with the Antarctic ice cap as the major freshwater reservoir. By comparison, all the other contributions are of a minor nature.

Available Water and Global Warming

The most important source of readily available, albeit recycled, fresh water is rain, the distribution of which is quite erratic. As a result of rainfall and percolation from the water table to the topsoil, the total volume of moisture in the soil is 25 000 km^3. Plants normally grow on what is considered to be 'dry' land, but this is a misnomer, because even desert sand contains up to 15% of water. It appears that plant growth requires extractable water; thus, an ordinary tree withdraws and transpires about 190 l per day. Groundwater is an increasingly important

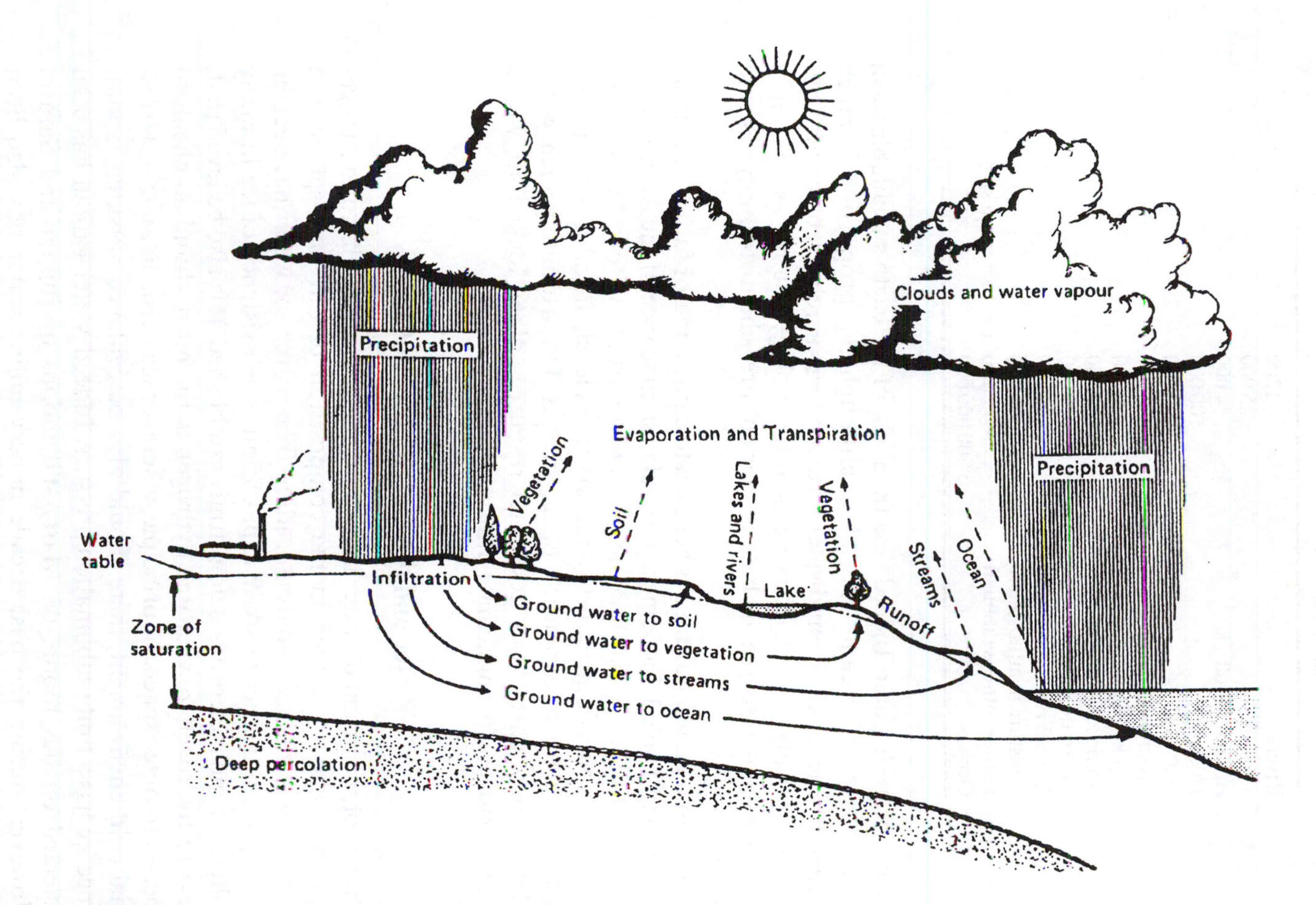

Figure 1.1 *Schematic representation of the hydrologic cycle*

Table 1.1 *Distribution and movement of earth's water resources (km^3)*

Rivers	1200	
Topsoil	25 000	
Annual runoff	34 000	
Lakes	100 000	
Glaciers	200 000	
Fresh groundwater (1 km depth)	4 000 000	
Arctic ice cap	3 000 000	
Antarctic ice cap	30 000 000	
Annual rainfall	2 000 000	
Moisture in atmosphere	60 000	
Photosynthesis (annual)	100	(= 10^{11} tonnes)
Oceans	1 300 000 000	

source of fresh water. Indeed, less than 3% of the earth's available fresh water occurs in streams and lakes, although the proportion is much higher in the UK. Groundwater hydrology is a relatively young, but rapidly growing, branch of science and technology, mainly as a result of the growth of research in botany, agriculture, chemistry, physics and meteorology.

Now that global warming is believed to cause a major future threat, the numbers in Table 1.1 assume a particular significance. If the Antarctic ice cap were to recede and become subject to partial melting, the water so produced would contribute to a rise in the sea level, in addition to any rise caused by thermal expansion of the oceans. Unfortunately such water which is now part of our freshwater resources, albeit locked up, would become useless for immediate utilisation.

Water and the Development of Life

Studies of the origin of water in the universe and the prehistoric changes that may have occurred in the composition of our atmosphere and hydrosphere make fascinating reading. The search for water has become an important aspect of space exploration. The existence of ice in many cold stars and meteorites is now firmly established. It is also believed that, next to hydrogen, oxygenated hydrogen is the most abundant chemical species in outer space. In principle, wherever ice exists in an extraterrestrial cold environment, there should also be evidence of water vapour, since ice has a finite sublimation pressure. Indeed, water vapour has been detected on our moon, on Mars, and moons of Jupiter and Saturn. However, during the past decade noncrystalline water has also been detected in the photosphere of the sun. By comparison with high-temperature emission spectra of very hot water, the infrared lines observed in sunspot spectra have been assigned to characteristic rotation and rotation–vibration transitions involving H_2O molecules and OH radicals. At ca. 3000 K the spectra correspond to approximately equal concentrations of molecules and radicals.

For any discussion of life on earth it is very important to establish when, and how, molecular oxygen first made its appearance. Early prokaryotes had minimal requirements of H, C, N, O, S and P, with a further selective management of Na, K, Mg, Ca, Fe, Mo, Se and Cl for energy regulation and electron transport. There is now little doubt that there existed enough H_2S to provide the reducing environment needed for the essential reactions with CO_2 to synthesise organic molecules. However, the low solubility of H_2S did not make it the ideal provider of hydrogen. Water itself was a much better provider of hydride. Eventually, with the aid of manganese as catalyst, living systems found the means of releasing hydride from water, but with a devastating side effect: the release of molecular oxygen which was the enemy of the reductive cell chemistry of primitive life. Eventually living organisms came to terms with oxygen and were able to gain energy from its breakdown:

$$O_2 + \text{C/H/N compounds} \rightarrow N_2 + CO_2 + \text{energy}$$

Different forms of simple organisms thus came to coexist: some remained anaerobic, while others made use of oxygen and became photosynthetic, while yet others became parasitic, using plants and oxygen as energy sources; they ultimately developed into animals. Free oxygen was produced by the splitting of water, first by high energy radiation and later also by photosynthesis. The original earth atmosphere is believed to have been composed of methane, ammonia, carbon dioxide, nitrogen and water vapour; it had a reducing character. It might also have included hydrogen and helium. An alternative view is that the present atmosphere is due more to the degassing of the earth's interior, e.g. by volcanic eruptions. Some planets still possess their 'original' atmospheres; in this context the current Galileo space probe exploration of Jupiter and its atmosphere takes on a special significance. Hopefully, we shall learn something about the origin of water in the solar system, and particularly about the development of our own atmosphere and hydrosphere. Only in this way will it become possible to relate the properties of water to the development of life processes on earth.

Carbon dioxide was first produced through the erosion and decomposition of minerals. In the presence of free oxygen a methane–ammonia atmosphere is unstable, methane being oxidised to water and carbon dioxide. The generation and consumption of oxygen are finely balanced:

the production through photosynthesis amounts to ca. 6.8×10^{15} mol a^{-1}, of which > 99% remains in the atmosphere. Of this, 90% is required for oxidative reactions that accompany the weathering and erosion of rocks. Thus only 3×10^9 mol oxygen remain for an enrichment of the atmosphere. By a combination of the above estimates with the known extension of plant and animal life on earth it is possible to sketch the increase of oxygen in the earth's atmosphere to reach its present value of 23% by weight. This is shown in Figure 1.2; the dramatic increase in the rate of oxygen production dates from the later Palaeozoic era, when a rapid growth of plant life took place which was then followed by a corresponding increase in animal life. Nevertheless, even at that time, life had already existed for three billion years.

Aerobic and Anaerobic Life Forms

Life began in hot water and was characterised especially by a rich diversity of algal species. The simplest prokaryotic forms of life can operate and reproduce with a rudimentary genetic machinery that requires no light. The terrestrial species exist at pH 1; they generate energy by oxidative reactions and thus they require not only sulfur, but also atmospheric oxygen. Species that exist in the submarine hot springs, on the other hand, are generally anaerobic. Apart from an absolute requirement for water, they also need variable amounts of CO_2, H_2, H_2S, CO, CH_4 and SO_4^{2-}. The nearest relatives of the hyperthermophiles are less heat tolerant (ca. 70 °C) and gain their energy by photosynthesis. The line

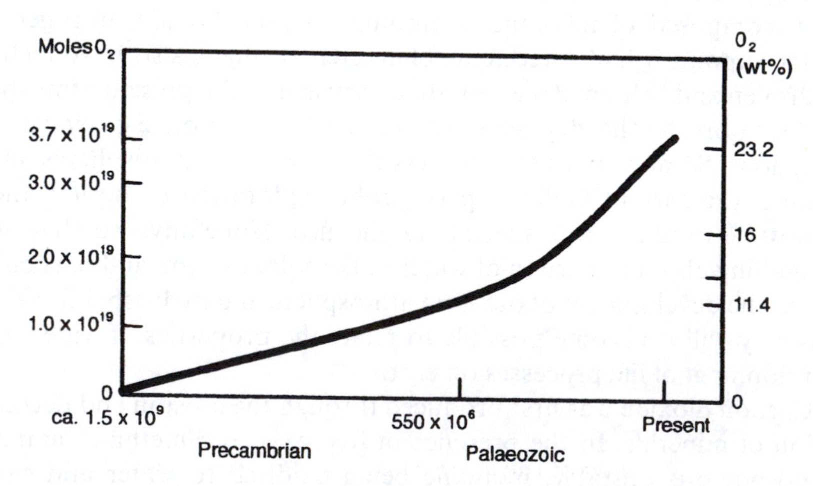

Figure 1.2 *Molecular oxygen in the evolving earth's atmosphere*

of development then leads to the microfungi and eukaryotes, which, although they exhibit increasing heat sensitivity (50–60 °C), can withstand temperatures that are still lethal for even the most primitive animals. As the earth cooled down, so evolutionary pressures led to the survival and further development of less heat tolerant organisms. We now talk of 'ambient temperatures' of the order of 0–40 °C and have reached the stage in the cooling down process, where scientists are becoming increasingly interested in the study of psychrophily, because cold has already become the most widespread and lethal enemy of life on earth.

Recent years have witnessed a growing interest by microbiologists, molecular biologists and biochemical engineers in extremophilic organisms, mainly directed towards a better understanding of protein stability under extreme conditions of temperature and pressure. The deep-sea trenches in the Pacific Ocean are a rich source of a variety of such microorganisms. Seen from the perspective of these so-called extremophiles, such a label is surely inappropriate. They would regard their habitat as the natural physiological environment. They would therefore classify any form of life that can survive and reproduce in an oxygen-containing atmosphere at 20 °C as a hyperextremophile!

Impact of the Physical Properties of Water on Terrestrial Ecology

The maintenance of life, as we know it, on this planet depends critically on an adequate supply of water of an acceptable quality and also on a well-regulated temperature and humidity environment. The oceans satisfy both these requirements, either directly of indirectly. They constitute our most important reservoirs of water and energy. They exert a profound influence on the terrestrial climate and on the levels of precipitation. A simple calculation serves to illustrate this stabilising effect:

The Gulf Stream is 150 km wide and 0.5 km deep and flows at a rate of 1.5 km^{-1} from the Gulf of Mexico to the Arctic Ocean, north of Norway. The temperature drop of the water during its passage north is about 20 °C. This temperature drop is equivalent to an energy transfer of about 5×10^{13} kJ km^{-3}, equivalent to the thermal energy generated by the combustion of 7 million tonnes of coal. The above dimensions, coupled with the rate of flow of the current provide for a movement of 100 km^3 of water per hour, equivalent therefore to the energy supplied by 175 million tonnes of coal. All the coal mined in the world in one year would be able to supply energy at this rate for only 12 h. The warm ocean currents thus act as vast heat exchangers and are responsible for maintaining a temperate climate over much of the earth's surface. The oceans are able to store such large amounts of energy by virtue of the large (for a substance with a molecular weight 18) heat capacity of water, one of its abnormal physical properties, to be discussed in more detail in several chapters of this book.

Other physical properties of liquid that greatly influence our ecological environment are the low density of ice relative to that of the liquid, and the phenomenon of the negative coefficient of expansion of cold water. Between them, these two properties are responsible for the freezing of water masses from the surface downwards, with obvious implications for the survival of aquatic life.

Chapter 2

Structure of the Water Molecule and the Nature of the Hydrogen Bond in Water

The Isolated Water Molecule

Ultimately the macroscopic behaviour of water depends on the details of its molecular structure. Quantum mechanics permits a theoretical analysis of the molecular structure from a knowledge of the masses, charges and spins of the subatomic particles involved. This requires the solution of the Schrödinger equation, the Hamiltonian for the water molecule being given by

$$H = E_n + E_e + U(\mathbf{r}, \mathbf{R}) \tag{2.1}$$

where the first two terms on the right-hand side represent the kinetic energy of the three nuclei and 10 electrons respectively, and where $U(\mathbf{r}, \mathbf{R})$ is the potential-energy function that contains the electrostatic interaction contributions from all pairs of particles, the particle coordinates being given by $\mathbf{r}$ and $\mathbf{R}$. Because of the large difference in the masses of the nuclei and the electrons, the first term in equation (2.1) contributes little to the total energy, and thus an approximate, but adequate, solution of the Schrödinger equation can be given by considering the electronic motions in a force field of fixed nuclei; this is known as the Born–Oppenheimer approximation. The solution of equation (2.1) then follows along conventional lines; electronic eigenvalues and eigenfunctions must first be defined. Although there is no exact method for doing this, various approximate treatments are available for estimating the upper limit of the electron energy. It is beyond the scope of this review to discuss molecular orbital calculations for complex molecules (water is a complex molecule in that it is polynuclear). A representation of the total calculated electron density of the ground state of water in the molecular plane is shown in

Figure 2.1. This graphically illustrates the asymmetry of the molecule. Although the excited electronic states are of interest for spectroscopy and the elucidation of reaction mechanisms, much less is known about them as regards theoretical predictions.

The solution of the nuclear Schrödinger equation provides information about the internal motions (vibration and rotation) of the molecule, and the theoretical results are consistent with the information provided by infrared spectroscopy. The equilibrium geometry of the isolated H_2O molecule indicates that the O–H bond length is 0.0958 nm and the H–O–H angle is 104°27′. The principal vibrations are shown in Figure 2.2. These frequencies are modified in the condensed states, where intermolecular effects become important.

The calculated electron density distribution surface shown in Figure 2.1 lends support to an earlier, empirical model for the water molecule, the Bjerrum four-point-charge model, which is illustrated in Figure 2.3. The oxygen atom is situated at the centre of a regular tetrahedron with

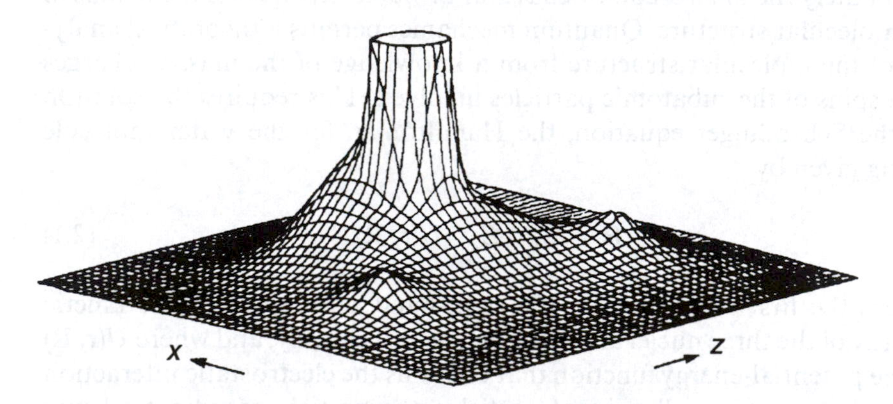

Figure 2.1 *A three-dimensional plot of the total electron density of the ground state of water. The viewer is oriented 45° clockwise from the line bisecting the HOH angle and 15° above the plane*
(Reproduced by permission from *Water – A Comprehensive Treatise*, ed. F. Franks, Vol. 1, Plenum Press, New York, 1971, p. 464)

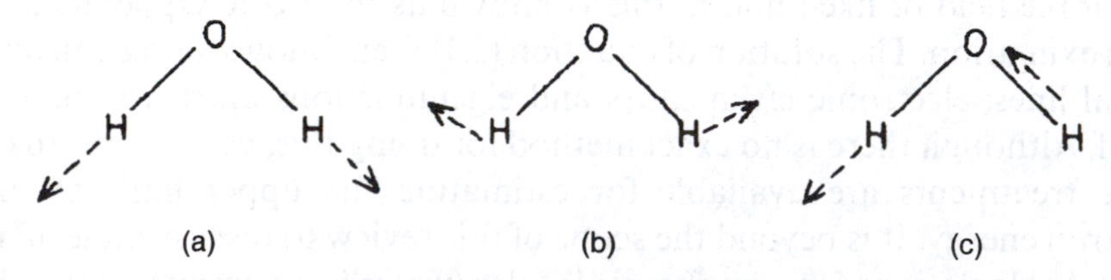

Figure 2.2 *The three principal vibration frequencies of the water molecule:* (a) *the symmetric valence stretching* (ν_1), (b) *the deformation mode* (ν_2), (c) *the asymmetric valence stretching* (ν_3)

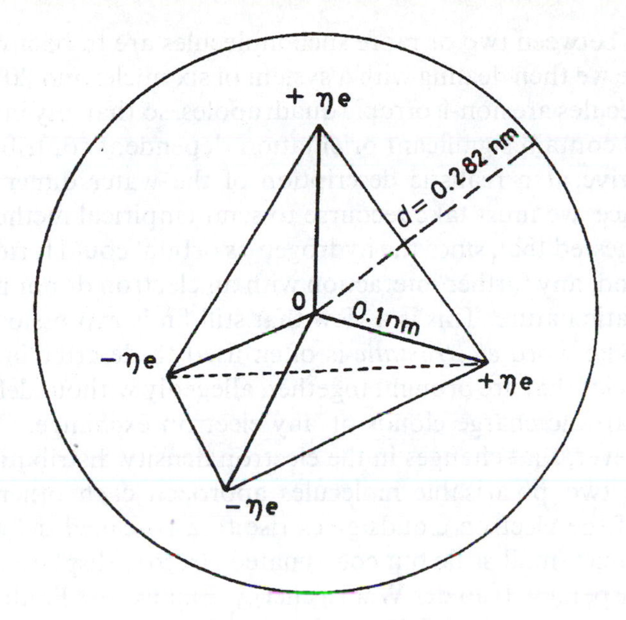

Figure 2.3 *The Bjerrum four-point-charge model for water*
(Reproduced by permission from *Water – A Comprehensive Treatise*, ed. F. Franks, Vol. 1, Plenum Press, New York, 1971, p. 417)

the fractions of charge $\pm\eta e$ placed at the vertices of the tetrahedron at distances 0.1 nm from the centre. The van der Waals diameter (d) assigned to this molecule is 0.282 nm, identical with that of neon (water and neon are isoelectronic). According to this representation the vertices carrying positive charge are the positions of the two hydrogen atoms, with the two lone electron pair orbitals carrying the negative charge, directed towards the other two vertices. It is easily seen that if two such molecules are allowed to approach one another, their interaction would have the characteristics of what has become known as the hydrogen bond, owing to electrostatic interactions between the charges.

Over the years, the original Bjerrum four-point-charge model has been subjected to several refinements to bring it more in line with the quantum mechanical results, according to which the molecule is not a regular tetrahedron. Rather, the O–H distance is longer than the distance between the centre and the position of negative charge. Nevertheless, the Bjerrum model has been remarkably successful in accounting, at least qualitatively, for the physical properties of ice and water.

The Water Dimer

If it is difficult to calculate *ab initio* the structure and properties of the isolated water molecules, then such difficulties are compounded when interactions between two or more such molecules are to be investigated. Not only are we then dealing with a system of six nuclei and 20 electrons, but the molecules are non-isotropic quadrupoles, so that any interactions are likely to contain significant orientation-dependent contributions. In order to arrive at a realistic description of the water dimer potential energy surface, we must take recourse to semi-empirical methods. Pauling first suggested that, since the hydrogen 1*s* orbital could form only one covalent bond, any further interaction with an electron donor must be of an electrostatic nature. This is a view that still finds expression in many textbooks. The word *electrostatic* is often used to describe interactions between species that are brought together, allegedly without deformation of their electronic charge clouds or any electron exchange. One might expect, however, that changes in the electron density distribution would occur when two polarisable molecules approach each other. A large distortion of the electron clouds gives rise to a so-called delocalisation energy, whereas small-scale but coordinated electron displacements give rise to the dispersion (van der Waals) energy. Finally, the Pauli exclusion principle dictates that for a full description of an interaction, a repulsive contribution must be included. It is now held that the hydrogen bond contains all the above contributions, although there is still some uncertainty about their correct relative weightings.

The quantity of greatest immediate interest is the pair correlation $g(r)$ which specifies the probability of finding an atom at a distance r from another atom placed at $r = 0$. Figure 2.4 shows the $g(r)$ values of argon (full-drawn curve) and the oxygen atoms in water (broken curve), both scaled to their respective molecular van der Waals radii. The peak positions indicate nearest neighbour distances, the peak widths the fluctuations from the equilibrium value and the peak areas provide an estimate of the number of neighbours within a given distance (coordination number n).

Leaving aside for the moment second and more distant peaks, even the nearest neighbour peak area for water points to one of its anomalous properties. The liquid argon coordination number, $n = 11$, is characteristic of a 'simple' substance, where melting of a close-packed crystal ($n = 12$) has led to a small volume expansion, accompanied by a small reduction in n. Water, by contrast, has $n = 4.4$ (for ice, $n = 4$), i.e. an *increase*, equivalent to a collapse (contraction) of the solid upon melting.

A consideration of the positions of oxygen atoms is not sufficient for a complete specification of the relative positions of two H_2O molecules in space. The corresponding intermolecular O–H and H–H pair correlation functions need to be specified. Bearing in mind the tetrahedral molecular structure of H_2O, such a complete specification requires the incorporation of orientation-dependent terms, so that the correlation function

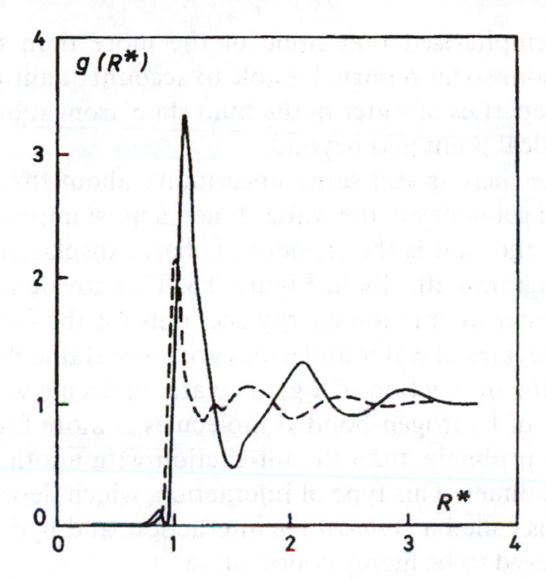

Figure 2.4 *Radial distribution function for oxygen atoms in water (broken line) at 277 K and for argon (solid line) at 84.25 K, both as functions of the reduced distance (see Chapter 5, p. 43)*

must be written as $g(r,\Omega)$, where Ω represents the angular coordinates (hydrogen bond donor and acceptor angles).

The correlation functions $g_{OH}(r)$ and $g_{HH}(r)$ are of necessity more complex in appearance than $g_{OO}(r)$; they are also more difficult to measure experimentally (see Chapter 5). On the other hand, a knowledge of $g_{OO}(r)$ alone does not provide sufficient information for the complete specification of the geometry of the water dimer.

Three approaches are in use for the measurement or calculation of $g(r)$. Experimental methods rely on X-ray and neutron scattering, while theoretical approaches assume a knowledge of the molecular pair potential function $u(r)$. Nearly all theoretical studies on water dimers have in the past relied on computer simulation approaches which allow structural and dynamic properties to be calculated from assumed potential functions.

Many methods have been proposed for the 'construction' of potential functions. It is beyond the scope of this review to discuss their relative merits except to state that there is general agreement on the nuclear dissociation energy (20–35 kJ) and on the equilibrium O–O separation (0.26–0.30 nm). There seems to be little difference in the total dimer energy for small (< 25°) amounts of hydrogen-bond bending, so that in liquid water the possibility of bent hydrogen-bond must be taken into account. Other features that are affected by hydrogen-bond bending include the O–O distance and the molecular dipole moment.

It must be emphasised that none of the more than twenty dimer potential functions so far reported is able to account quantitatively for all the physical properties of water in the fluid state, from subzero temperatures to the critical point and beyond.

Finally, while there is still some uncertainty about the details of the hydrogen-bond geometry in the water dimer, a most important feature is beyond doubt, and that is the element of charge displacement, which is represented diagrammatically in Figure 2.5. This covalent contribution to the water–water interaction energy accounts for the fact that trimers and larger aggregates of water molecules are more stable than the simple dimer; that is, the interaction of a given water molecule with an already existing cluster of hydrogen-bonded molecules is more favourable, and therefore more probable, than the interaction with another single molecule to form a dimer. This type of interaction, which depends on previous processes, is called a *cooperative* interaction, and hydrogen-bonded in water is believed to be highly cooperative.

A physical manifestation of the hydrogen bond becomes apparent from a comparison of the infrared and Raman spectra of water vapour with those of liquid water and ice, another hydrogen-bonded form of water. The normal modes characteristic of the isolated water molecule, shown in Figure 2.2, are markedly perturbed when the vibrating O–H bond 'senses' another water molecule oriented in such a way that its lone pair electrons (the negatively charged vertex of the tetrahedron in Figure 2.3) face the hydrogen atom of the vibrating bond. Thus, the frequency of the ν_1 vibration, the symmetric O–H valence stretching, decreases from 3707 cm^{-1} in the isolated molecule to 3628 cm^{-1} in liquid water and to 3277 cm^{-1} in ice. In addition, new spectral features appear, because the hydrogen bond O–H···O itself has additional vibrational degrees of freedom. Since hydrogen bonds are weak compared to the normal covalent bond, these new lines appear at the low-frequency end of the water spectrum, $< 1000\ cm^{-1}$.

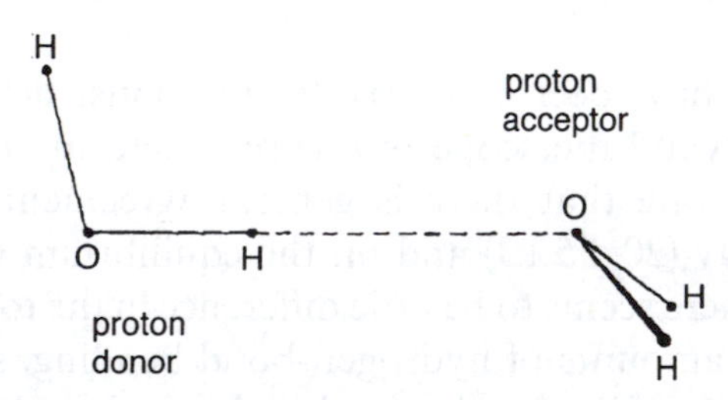

Figure 2.5 *Hydrogen bond geometry of the water dimer, showing the contribution from charge displacement*

Chapter 3

Physical Properties of Liquid Water

The Place of Water in a General Classification of Liquids

There are several problems inherent in the formulation of a molecular description of a liquid. On the one hand one can begin by considering a dense, crystalline solid in which atoms or molecules occupy given positions in space and where the only possible motion is oscillatory, centred about the equilibrium positions. A liquid can then be regarded as a perturbed solid in which the degree of order has been reduced upon melting, emphasis still being placed on the positions occupied by molecules over a time average.

On the other hand, one can proceed from a consideration of the dilute gas in which molecules are free to move in a random manner and do not interact with one another; that is, they do not take up preferred positions relative to each other. Starting from this concept, a liquid is then regarded as a dense gas in which diffusive motion is still important, albeit inhibited by the proximity of other molecules and subject to a considerable degree of molecular interaction. Both approaches have their merits, although the development of X-ray and neutron diffraction techniques has placed the emphasis on the solid-like features (i.e. structure) of liquids.

Liquids can be classified according to the nature of the intermolecular forces involved, as shown in Table 3.1. These interactions will determine the appearance of the (P–V–T) phase coexistence diagram which is therefore characteristic for a given substance. Figure 3.1(a) shows such a P–V–T surface for a given mass of substance, and Figure 3.1(b) is the P–T projection of this surface, a more familiar representation of the phase diagram. The solid/fluid transition is accompanied by discontinuities in many physical properties, e.g. density, internal energy, specific heat, refractive index and dielectric permittivity. At high pressures the differences in the properties characteristic of the solid and the fluid remain

Table 3.1 *Intermolecular forces in various types of liquids*

Nature of molecule	*Predominating intermolecular force*
Monatomic (spherical)	Van der Waals (dispersion), short range
Homonuclear diatomic	Van der Waals (dispersion), short range + quadrupole interactions
Metals, fused salts	Electrostatic, long range
Polar	Electric dipole moments
Associated	Hydrogen bonds, orientation dependent
Macromolecules	Some of the above, depending on chemical nature; also intramolecular forces
Helium	Quantum effects

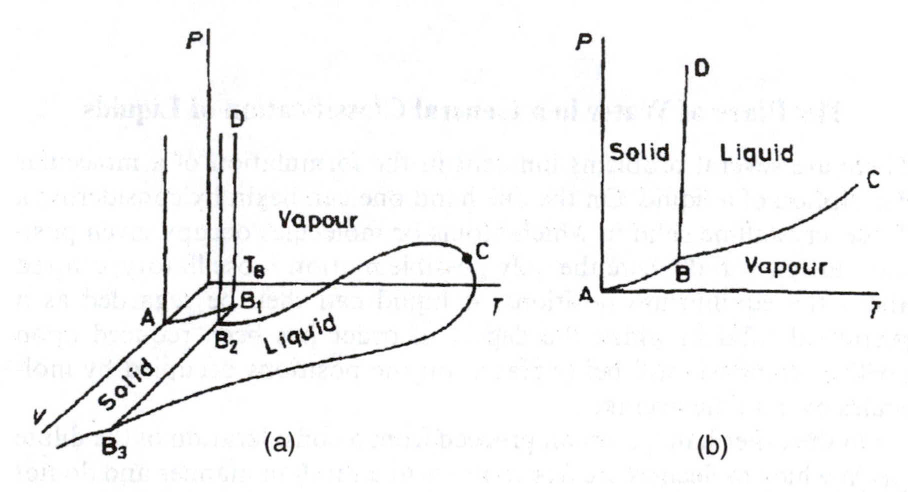

Figure 3.1 (a) *General* P–V–T *surface for a given mass of substance.* T_B *is the triple point and C the critical point;* (b) *the* P–T *projection of the surface*

unchanged, whereas those of the liquid and the gas decrease and merge at the critical point.

Figure 3.2 is a *P–T* projection of the high-pressure portion of the phase diagram for water. In the supercritical region at a given pressure, density and temperature can be changed independently. For instance, at 500 °C, water can be compressed to a density of 2.1×10^3 kg m^{-3}, where it solidifies to ice VII. High-pressure phase data are of great importance in the design of power stations, the control of corrosion and in geochemistry.

Some physical properties of typical members of the groups of substances shown in Table 3.1 are summarised in Table 3.2. Even a cursory examination illustrates the anomalous position of water. The volume contraction that accompanies the melting of ice is well known, and while it is not unique it is nevertheless rare. The very large liquid temperature

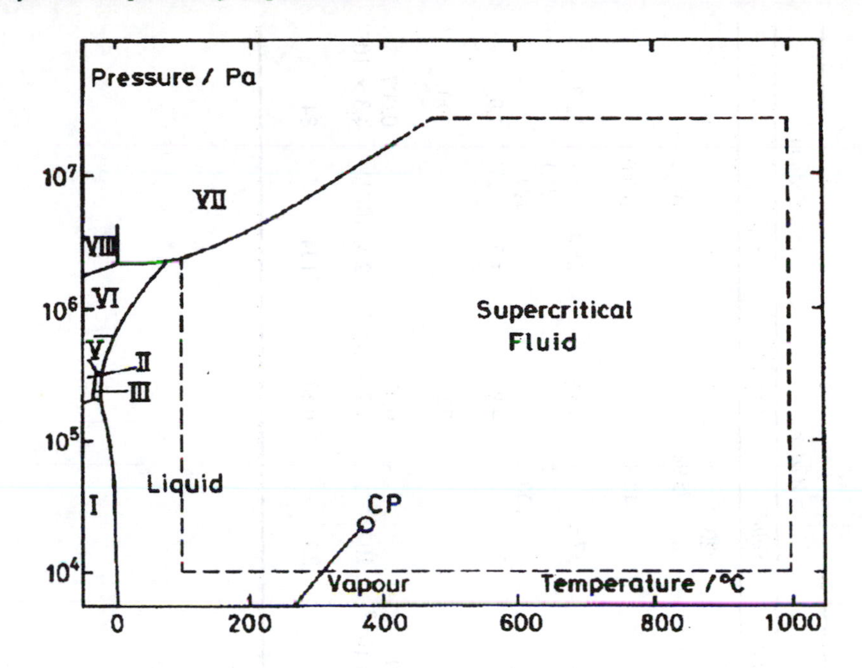

Figure 3.2 *The* P–T *projection of the high-pressure portion of the phase diagram of water. CP is the critical point. The various ice polymorphs are indicated by Roman numerals (see Chapter 4 for details)*

range is an indication of long-range forces in liquid water. This is confirmed by the very small latent heat of fusion which only amounts to 15% of the latent heat of evaporation and suggests that liquid water retains much of the order of the solid state and that this order is only destroyed at the boiling point. In view of the low-density crystal structure of ice, it is remarkable that the compressibility of liquid water is so low, because compressibility is a measure of the repulsive intermolecular forces.

Probably the most anomalous property of liquid water is its large heat capacity which is reduced to half its value upon freezing or boiling. Like several other properties of water, this extraordinarily large heat capacity plays an important role in maintaining a climatic environment in which life, as we know it, can exist. The phenomenon has already been described by way of the Gulf Stream example.

While the *P–V–T* behaviour of water, and especially of liquid water, may not be typical of a small molecule, its transport properties show fewer eccentricities. Although the viscosity is slightly higher than might be expected, the orders of magnitude of the coefficients of viscosity, self-diffusion and thermal conductivity are those typical of molecular liquids, suggesting that the transport mechanisms underlying these phenomena

Table 3.2 *Physical properties of some typical liquids*

	Argon		*Benzene*		*Water*		*Sodium*	
	Solid	*Liquid*	*Solid*	*Liquid*	*Solid*	*Liquid*	*Solid*	*Liquid*
Density (kg m^{-3})	1636	1407	1000	899	920	997	951	927
Latent heat of fusion (kJ mol^{-1})	7.86		34.7		5.98		109.5	
Latent heat of evaporation (kJ mol^{-1})	6.69		2.5		40.5		107.0	
Heat capacity [J (mol K)$^{-1}$]	25.9	22.6	11.3	13.0	37.6	75.2	28.4	32.3
Melting point (K)	84.1		278.8		273.2		371.1	
Liquid range (K)	3.5		75		100		794	
Isothermal compressibility [(N m^2)$^{-1}$]	1	20	8.1	8.7	2	4.9	1.7	1.9
Surface tension (mJ m^{-2})		13		28.9		72		190
Viscosity [(Poise = 10^{-1} kg m^{-1} s^{-1})]		0.003		0.009		0.01		0.007
Self-diffusion coefficient (m^2 s^{-1})	10^{-13}	1.6×10^{-9}	10^{-13}	1.7×10^{-9}	10^{-14}	2.2×10^{-9}	2×10^{-11}	4.3×10^{-9}
Thermal conductivity [J (s m K)$^{-1}$]	0.3	0.12	0.27	0.15	2.1	0.58	134	84

are common for argon, benzene and water, if not sodium. Some salient physical properties of liquid water will be described in later sections.

Isotopic Composition

The physical properties of water listed in the various compilations refer to mixtures of varying isotopic compositions. On a molar basis such properties do not vary much, but since engineering work usually requires physical data on a weight basis, allowance must be made for isotopic variability. There exist two stable hydrogen and three stable oxygen isotopes, as well as the radioactive ^3H with a half-life of 12.6 a whose presence in natural waters depends on the frequency of thermonuclear testing but is negligible for purposes of calculating physical properties. It is, however, very important for purposes of dating underground water reservoirs. Tables 3.3 and 3.4 summarise the natural abundance of the various isotopes and the major molecular species. Also, since water contains low concentrations of ions (H_3O^+ and OH^-), second-order complexities arise concerning relative magnitudes of the ionisation constants of the various isotopic forms.

Although natural waters are subject to small variations in isotopic composition, such variations are insignificant compared with the alterations caused during the purification of water in the laboratory. Thus, the density difference between the first and last fractions of a distillation can be 20 ppm.

Table 3.5 shows that wide divergences can exist in certain properties between molecules that are physically and chemically very similar indeed.

Table 3.3 *Isotopic masses and abundances of water elements*

Isotope	*Atomic mass*	*Natural abundance (wt%)*
^1H	1.007825	99.970
^2H	2.01410	0.030
^3H	3.01605	0
^{16}O	15.99491	99.732
^{17}O	16.9914	0.039
^{18}O	17.99916	0.229

Table 3.4 *Natural abundance of isotopic water species (ppm)*

$^1H_2{}^{16}O$	$^1H_2{}^{18}O$	$^1H_2{}^{17}O$	$^1H^2H^{16}O$
997 280	2000	400	320

Table 3.5 *Maximum density of isotopic modifications of water*

Property	$^{1}H_2O$	$^{2}H_2O$	$^{1}H_2{}^{18}O$	$^{2}H_2{}^{18}O$	$^{3}H_2O$
Maximum density (kg m^{-3})	999.972	1106.00	1112.49	1216.88	1215.01
Temperature of maximum density (°C)	3.984	11.185	4.211	11.438	13.403

The temperature of maximum density is the most striking example, but other thermodynamic properties also show these differences, e.g. heat capacity, volume and compressibility. All these differences are of importance in calculations of equations of state suitable for engineering applications and for the development of international standards.

Thermodynamic Properties

Even the engineers of remotest antiquity were able to provide water of adequate quality for their cities. Water engineering flourished during the Roman Empire, with many aqueducts built at the time still surviving. In the Middle Ages the use of water mills flourished in Europe. Leonardo da Vinci wrote a monograph entitled *Del moto e misura dell'acqua* in which he catalogued many of the important properties of the liquid. Practical knowledge of the properties of water – density, volatility, and viscosity – thus preceded the development of scientific understanding. Similarly, Watt patented the condenser in 1769, making the steam engine a practical proposition some time before the discovery of the compound nature of water, shown by Cavendish in 1781, and well before the development of thermodynamics during the nineteenth century.

Today, the properties of water play a key role in, for example, oceanography and limnology, hydraulics, biochemistry and physical chemistry. Usually, it is the thermodynamic and transport properties that are of the greatest practical importance. For many applications, values of properties are needed with high precision, even though their relation to structure on a molecular scale may still be imperfectly understood. Accordingly, many measurements and tabulations have been made for particular purposes, the most noteworthy being steam tables for the design of power plants.

The thermal power industry needs the design tables to remain unchanged for long periods. Political considerations are also said to influence the method of arriving at the values and tolerances of the skeleton table. Thus, steam tables should not be considered critical compilations for other purposes and, in particular, they should be differentiated only with caution. Their only utility for others is that they tabulate properties over a wide range of conditions and are available in various engineering units.

Classical thermodynamics is based on laws of bulk matter, laws that are independent of the structure of matter. Hence thermodynamic measurements have a permanent value, independent of dispute over the detailed structure of the liquid. They do, however, often play a central role in the development of models, even when such models are not based solely on thermodynamic information. One might even go so far as to claim that structural models should never be based solely on thermodynamic information.

The Gibbs free energy G is here chosen as the primary thermodynamic function from which other equilibrium properties are obtained by well-known standard relations:

Volume	$V = (\partial G/\partial P)_T$
Entropy	$S = -(\partial G/\partial T)_P$
Enthalpy	$H = G + TS$
Internal energy	$U = H - PV$
Isobaric thermal expansivity	$\alpha_P = [\partial(\ln V)/\partial T]_P$
Isothermal compressibility	$\kappa_T = -[\partial(\ln V)/\partial P]_T$
Isopiestic (adiabatic) heat capacity	$C_p = (\partial H/\partial T)_P$
Isochoric heat capacity	$C_v = (\partial U/\partial T)_V$

Volume Properties

The density is important for most other studies of water. It has therefore been the subject of numerous measurements, especially in the temperature range 0–40 °C, and many extensive tabulations have been compiled.

Like the volume at atmospheric pressure, the isothermal compressibility displays a minimum, in this case near 46.5 °C. The favourite method for the measurement of κ_T is the velocity of sound, which can be determined with a high degree of precision. It should not be inferred that this particular temperature is of some deep significance for the properties of water, since simple transformations give other functions of equivalent physical content with extrema at other temperatures. Thus, for the velocity of sound u itself, i.e. for $(\partial \rho/\partial P)_S$, the extremum is at 74 °C, for the isentropic compressibility, $\kappa_S = -[\partial(\ln V)/\partial P]_S$, it is 64 °C, and for $(\partial V/\partial P)_T$ it is 42.3 °C, all of these differing from the 46.5 °C for κ_T. The existence of these temperatures is related to the maximum-density phenomenon, but it is clear that there is no single temperature above which water behaves as a *normal* liquid and below which it is peculiar.

At atmospheric pressure, the temperature of maximum-density, that is, the temperature at which

$$(\partial\rho/\partial T)_P = 0$$

is 3.98 °C. At higher pressures, the maximum density moves to lower temperatures, with the dependence given by

$$\left(\frac{\partial T}{\partial P}\right)_{\rho\ \max} = -\frac{[(\partial/\partial T)(\partial\rho/\partial P)_T]_P}{(\partial^2\rho/\partial T^2)_P}$$

This is evaluated from the temperature dependence of the compressibility and of the thermal expansion, and it is found that $(\partial T/\partial P)_{\rho\ \max} = -0.0200 \pm 0.0003$ deg bar^{-1}. This line reaches the ice–liquid equilibrium line at -4 °C and 400 bar. For D_2O the corresponding value is -0.0178 deg bar^{-1}. The position of this line is a measure of the peculiarity of liquid water, and with D_2O, not only is it higher at atmospheric pressure, 11.44 °C versus 3.98 °C, but it does not move to lower temperatures as quickly with increasing pressure.

Figure 3.3 shows the stable region at temperatures up to the critical point. Three general areas may be distinguished. At lower pressures, near the critical point, the behaviour is essentially like that of other materials. At high pressures, e.g. at 10 kbar, the behaviour is also very regular. It is at low temperatures and pressures that the peculiar features of water, such as the minimum compressibility and maximum density, are found. At these lower pressures, the liquid can accommodate the larger volume entailed by less deformation of the intermolecular angles.

The most striking feature of the *P–V–T* properties in the low-pressure region is not apparent from data on the liquid alone, but is revealed by comparison with the ices. Liquid water is 10% denser than ordinary ice Ih, which means that the O–O distance is shorter. However, X-ray diffraction and the infrared spectrum indicate that the nearest neighbour O–O distance is longer in the liquid. Hence, the distances to higher neighbours must be shorter, and this is probably the case for second neighbours, implying deformation of O–O–O angles. It is the balance between first-neighbour distances increasing with temperature, which would increase the volume, and angular deformation increasing with temperature, which would decrease the volume, that produces the maximum density and the minimum compressibility.

Many equations have been proposed to represent the isothermal behaviour of liquid water. The observations can be represented reasonably closely by any of a number of equations with three parameters, one of which is usually the volume at low pressure. One such equation is

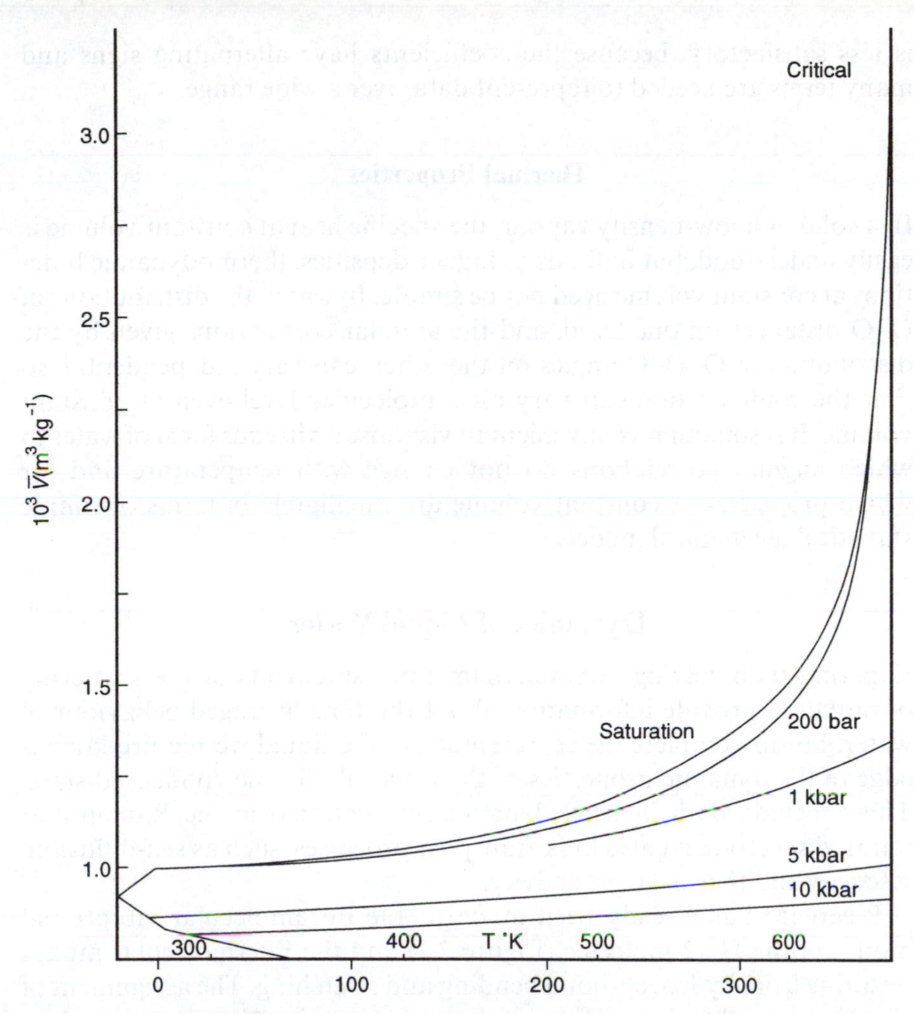

Figure 3.3 *Volume of liquid H_2O as a function of temperature. The field ends at the lower left at the freezing curves of the various ice polymorphs. Water is only strikingly different from other liquids at low temperatures and pressures*
(Reproduced by permission from *Water – A Comprehensive Treatise*, ed. F. Franks, Vol. 1, Plenum Press, New York, 1972, p. 381)

$$\frac{V_0 - V}{V_0 P} = \frac{A}{B + P}$$

where V_0 is the volume at zero pressure (sometimes approximated by the volume at atmospheric pressure), and A and B are adjusted to fit the data.

The representation of the isothermal equation of state by a power series in the pressure,

$$V = V_0 + a_1 P + a_2 P^2 + \dots$$

is less satisfactory, because the coefficients have alternating signs and many terms are needed to represent data over a wide range.

Thermal Properties

In a solid or a low-density vapour, the specific heat at constant volume is easily understood, but in fluids at higher densities, thermodynamic functions at constant volume need not be simple. In water, the distributions of O–O distances on one hand, and the angular correlations given by the distribution of O–O–O angles on the other, can vary independently, so that the configuration can vary on a molecular level even at constant volume. It is sometimes convenient to visualise a vitreous form of water in which angular correlations do not change with temperature and for which properties at constant volume are intelligible in terms of simple statistical mechanical models.

Dynamics of Liquid Water

Experiments involving thermodynamic measurements or the scattering of radiation provide information about the time-averaged behaviour of water, but to complete the representation of a liquid we require knowledge of the dynamic properties of the molecules in the condensed state. These include both internal dynamics, as reflected in the Raman and infrared spectra, and also bulk transport processes, such as self-diffusion, viscosity and thermal conductivity.

Reference has already been made to the intramolecular vibrational modes of the H_2O molecule (Figure 2.2) and the intermolecular modes resulting from hydrogen-bond bending and stretching. The assignment of specific spectral features to particular modes of vibration in the condensed state is complicated by coupling effects, since the vibrations of a given bond (e.g. the O–H stretch) are affected by the other O–H group in the same molecule and also by intermolecular effects involving O–H groups in neighbouring molecules. The interference of coupling effects has been circumvented by the investigation of spectra of a dilute solution of H_2O in D_2O, or vice versa. In that case the O–H (or O–D) groups are sufficiently separated so that coupling cannot occur.

Figure 3.4 shows the uncoupled Raman contours of the symmetric stretch (ν_1) vibration of O–D in a 11 mol% solution of D_2O in H_2O, as functions of temperature and pressure. Two features are of special interest:

(1) the contour is not symmetrical but exhibits a shoulder on the high frequency side of the main peak, and

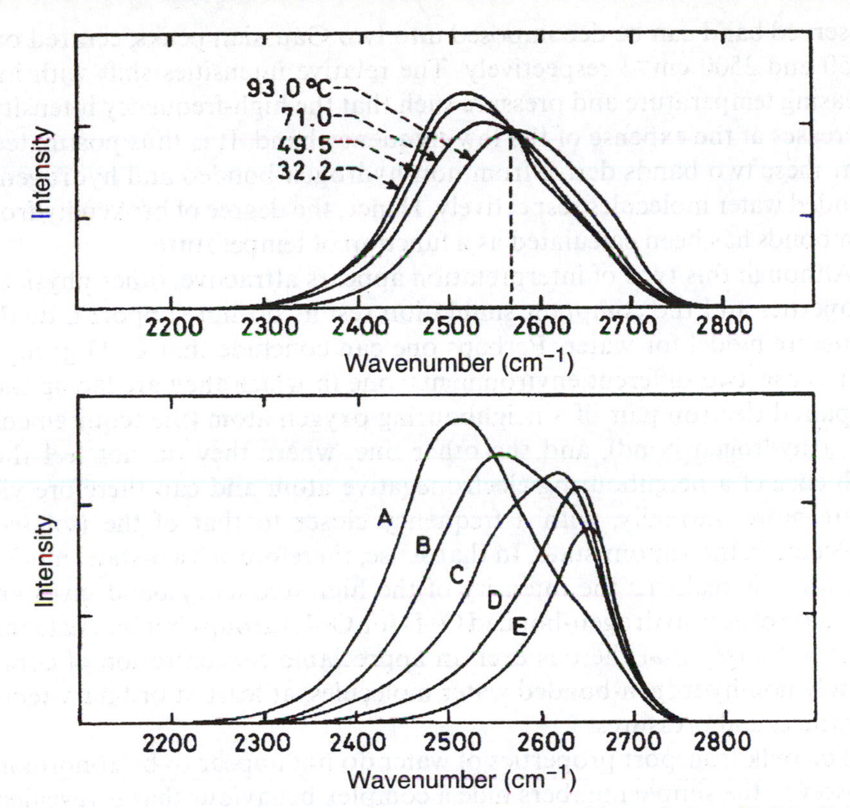

Figure 3.4 *The symmetric stretch vibration (ν_1 of O–D in a dilute solution of D_2O in H_2O as function of temperature and pressure. A: 20 °C, 100 × 105 MPa; B: 100 °C, 1 × 10^8 MPa; C: 200 °C, 2.8 × 10^8 MPa; D: 300 °C, 4.7 × 10^8 MPa; E: 400 °C, 3.9 × 10^8 MPa. Spectra a–d correspond to a constant density of 1000 kg m^{-3}*
(Reproduced by permission from *Water – A Comprehensive Treatise*, ed. F. Franks, Vol. 1, Plenum Press, New York, 1971, pp. 181, 182)

(2) the position of the main peak shifts to a higher frequency with increasing temperature and/or pressure.

In the high-temperature/pressure limit this frequency appears to be close to 2650 cm^{-1}. Since under these conditions water is well above its critical point, where presumably all hydrogen bonds have been broken, this frequency might be regarded as representing the vibration of non-hydrogen-bonded (free?) O–D bonds. However, in the vapour state the O–D vibrational frequency is 2727 cm^{-1}, so that even in the supercritical fluid the valence bond vibrations are perturbed by the presence of neighbouring molecules.

The asymmetry in the ν_1 band (also observed in the ν_2 and ν_3 bands) has been interpreted in terms of a dual-structure model for water. The

observed band can be decomposed into two Gaussian peaks, centred on 2650 and 2500 cm^{-1} respectively. The relative intensities shift with increasing temperature and pressure such that the high-frequency intensity increases at the expense of the low-frequency band. It is thus postulated that these two bands derive from non-hydrogen-bonded and hydrogen-bonded water molecules respectively. Hence, the degree of broken hydrogen bonds has been calculated as a function of temperature.

Although this type of interpretation appears attractive, other physical properties and the computer simulation results do not support a dual-structure model for water. Perhaps one can conclude that O–D groups can 'sense' two different environments: one in which they are facing the unpaired electron pair of a neighbouring oxygen atom (the requirement for a hydrogen bond), and the other one, where they do not feel the influence of a neighbouring electronegative atom and can therefore vibrate more normally, with a frequency closer to that of the isolated molecule in the vapour state. In that sense, therefore, a 'two-state' model for water is realistic; the intensity of the high frequency band gives an estimate of non-hydrogen-bonded O–D (or O–H) groups but its existence does not imply that there is ever an appreciable concentration of completely non-hydrogen-bonded water molecules, at least at ordinary temperatures and pressures.

The bulk transport properties of water do not appear to be abnormal. However, the simple numbers hide a complex behaviour that is revealed in the temperature and pressure dependence of the transport properties. The details of molecular motions in liquids can be expressed in terms of so-called time correlation functions, defined as

$$\Gamma(t) = < A(0)).A(t) >_0$$

$\Gamma(t)$ measures how long some property A persists before it is averaged out by random motions, collisions etc. $A(0)$ is the property under study at $t = 0$ and $A(t)$ is the same property after a time t. The brackets indicate that the average is taken over all molecules; for thermal equilibrium this average usually corresponds to a Boltzmann distribution. $\Gamma(t)$ is an *auto*correlation function, because it describes the time dependence of the motions of a particular molecule. Another useful function is the *cross*correlation function which describes correlated properties of a pair of molecules, such as the exchange of spins or the orientation of molecular dipoles, both of which persist for only short periods and are destroyed by random thermal motions. The description of molecular properties by means of time correlation functions is useful for liquids where there is an absence of long-range order because of random motion, but where intermolecular forces, although weak, oppose the complete randomisation of molecular properties.

Often $\Gamma(t)$ can be expressed in terms of a single exponential decay function and it is possible to define a *correlation* time τ_c which is given by

$$\tau_c = \int_0^\infty \Gamma(t)\mathrm{d}t$$

where τ_c is the reciprocal of the exponent $\exp(-t/\tau_c)$. In other words, τ_c is the time after which the magnitude of the property A has decayed to 1/e of its magnitude at $t = 0$, as a result of the various randomising influences.

In order to apply the concept of correlation functions to the study of molecular motion, let us consider the response of a dipolar fluid such as water to an external electric field. Assuming a simple exponential decay when the field is removed at $t = 0$, the time dependence of the molecular polarisation, $P(t)$, is given by

$$P(t) = P_0\exp(-t/\tau_c) \tag{3.1}$$

where τ_c is once again the correlation time, but is usually called the relaxation time.

If, instead of a stationary electric field, an oscillating field of angular frequency ω is applied, then the molecular dipoles have to reorient continuously. The ease of reorientation is related to the viscosity of the medium and the mobility of the electron clouds. The ease of polarisation is expressed in terms of the dielectric permittivity, ε, which is a complex quantity, given by

$$\varepsilon = \varepsilon' + j\varepsilon''$$

Figure 3.5 shows the frequency dependence of ε' and ε''; both these quantities can be expressed in terms of ω, τ_c, ε_s and ε_∞, where ε_s and ε_∞ are the values of ε' in the low- and high-frequency limit respectively. The value of ω at which ε' has fallen by half of its total decrease should be equal to τ_c^{-1}, if equation (3.1) correctly expresses $P(t)$, and for water this is indeed the case. In terms of molecular motions τ_c is a measure of the time taken by a molecule to perform one rotation, i.e. it is a measure of the rotational diffusion period. The Debye relaxation theory relates τ_c to the viscosity of the liquid:

$$\tau_c = \frac{8\pi\eta a^3}{2kT} \tag{3.2}$$

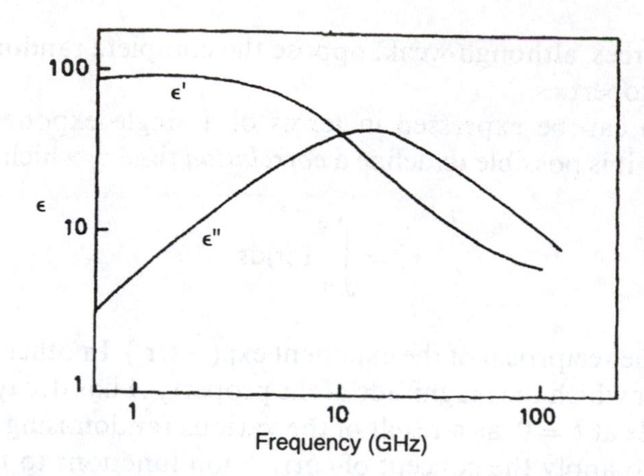

Figure 3.5 *Logarithmic plot of ε' and ε'' of water at 20 °C as a function of frequency*

where a is the molecular radius. For water at 25 °C measurements of ε provide a good estimate of $\tau_c = 8$ ps, which, when introduced into equation (3.2), gives $a = 0.14$ nm, in good agreement with the value obtained by taking one-half the inter-oxygen distance in ice. Rotational diffusion can also be studied by nuclear magnetic relaxation measurements, in which case the correlation function measures the decay of magnetisation. The estimate of τ_c is in good agreement with that obtained from dielectric measurements.

Knowledge of the correlation time τ_c also provides an estimate of the translational diffusion (self-diffusion) behaviour through the relationship

$$\tau_c = \frac{<r^2>}{12D^*} \tag{3.3}$$

which, in turn, is related to the viscosity through the Stokes–Einstein equation

$$D^* = \frac{kT}{6\pi\eta a} \tag{3.4}$$

Here D^* is the self-diffusion coefficient and $<r^2>$ is the mean square displacement of a molecule during the time τ_c. Independent estimates of D^* can be obtained by means of diffusion measurements involving isotopes (^2H, ^3H or ^{18}O) in water. All such measurements suggest that for H_2O at 25 °C, $D^* = 2.5 \times 10^{-9}\ m^2\ s^{-1}$ which is again consistent with the dielectric and NMR results.

At first sight such self-consistency and agreement between results obtained from different techniques is gratifying, but the actual results raise questions as to the mechanism of molecular transport in water. Thus, the theory of dielectric relaxation and rotational diffusion takes as its starting point the diffusion of an isotropic sphere in a viscous continuum. The theory therefore does not apply to liquids the molecules of which are associated by hydrogen bonds. Indeed, the results from dielectric measurements on alcohols or amines do not fit equation (3.2), because such liquids are believed to contain linear, branched chain and/or cyclic aggregates. It is surprising, therefore, that water, which is probably the most associated liquid, obeys this simple relationship which is based on a molecular model that can hardly represent the state of affairs in liquid water.

Again, self-diffusion measurements using ^3H and ^{18}O give the same result for D^*, which is moreover consistent with the value of a obtained by equation (3.2). This is also surprising, in view of the different types of motion that might be ascribed to oxygen and hydrogen in liquid water. Measurements with oxygen tracers might be expected to reflect more correctly the diffusion of molecules, because the oxygen atom in water contains most of the molecular mass and can only move by some more or less regular diffusion mechanism. The hydrogen atom, on the other hand, is expected to move primarily along the oxygen–oxygen axes (the hydrogen bonds), and such movement should be considerably facilitated compared with transport by random diffusion. This facilitated diffusion is most strikingly demonstrated in the proton conductance of ice which is higher by several orders of magnitude than that of liquid water.

While on the subject of the molecular transport of water, there is an anomaly that demonstrates the unique nature of liquid water; the deuterium isotope effect on τ_c is abnormally large. Thus $\tau_c(D_2O)/\tau_c(H_2O) = 1.3$ at 20 °C, decreasing with rise in temperature. Such an effect cannot be accounted for by the larger reduced mass of the D_2O molecule.

Bulk Transport Properties

Macroscopic hydrodynamics are based on the laws of bulk matter, such as the conservation of momentum, on the same level as those employed by classical thermodynamics. Here we consider the coefficients in the linear approximation. The (dynamic or shear) viscosity η measures the transport of momentum, the thermal conductivity λ measures the transport of heat, and the diffusivity, $\mathcal{D}$ the transport of mass.

Heat may be conducted by several mechanisms, for example by the diffusion of species, by phonons (elastic waves) and by photons. In nonmetallic solids at low temperatures, the principal method is by phonons. Photons become important at high temperatures. In the gas phase, heat is conducted primarily by the movement of molecules between collisions. In the liquid, scattering of phonons is large, rendering the phonon a less useful concept, and both collisional and diffusional processes are important.

In dilute vapours, viscosity and thermal conductivity are nearly proportional and show the same temperature dependence. In liquids, the temperature dependences are quite different. At constant pressure, viscosity decreases nearly exponentially with increasing temperature:

$$\eta = \eta_o \exp(-\Delta E_\eta^*/RT) \quad (3.5)$$

though for water ΔE_η^*, the Arrhenius energy of activation for viscous flow (approximately 21 kJ mol^{-1} at 0 °C), changes more with temperature than is the case with most liquids. Similarly, the pressure dependence at constant temperature can be represented by

$$\eta = \mathrm{A}' \exp(P\Delta V_\eta^*/RT)$$

where the phenomenological volume of activation ΔV_η^* is positive if the viscosity rises with increasing pressure and negative if it falls. In water, it is negative at low temperatures and positive at high temperatures.

In contrast, the thermal conductivity, λ, of liquids is nearly linear with temperature and, over narrow ranges, is usually represented by

$$\lambda = \lambda_o(1 + at)$$

These differences in temperature dependence suggest that there are at least two mechanisms effective in the liquid: rotational motion is more important for viscosity, but translational processes are more important for heat transport.

For practical purposes, equation (3.5) does not represent the viscosity of water with sufficient precision because $-\Delta E_\eta^*$ is taken as constant. The viscosity is often represented by the VTF (Vogel–Tammann–Fulcher) equation

$$\log \eta = \frac{A + B}{(T + C)} \quad (3.6)$$

where $T = -C$ is the glass transition temperature, T_g, which is estimated from equation (3.6) as 134 K. Even the VTF equation does not represent

$\eta(T)$ adequately for many purposes and has therefore been extended and modified.

The theory of bulk viscous flow is based on the picture that in a liquid holes exist, into which molecules can diffuse. It is thus assumed that a certain amount of free volume (volume not actually occupied by the molecules) exists in the bulk of the liquid and that transport on a molecular scale can only occur by utilising this free volume. An increase in pressure would be expected to hinder the diffusion process, since it must reduce the available free volume. Thus the viscosity of most fluids rises rapidly with increasing pressure, but the pressure dependence of the viscosity of water is different, as shown in Figure 3.6. At lower temperatures $\partial\eta/\partial P$ is negative at low pressures, passes through zero and becomes increasingly positive with increasing pressure. Above 33.5 °C the behaviour changes and the viscosity isotherms show a positive pressure dependence at all pressures. Clearly the negative pressure coefficient is related to the four-coordinated network of hydrogen-bonded molecules which dominates most physical properties of water at low temperatures, but no attempts at a quantitative interpretation of the viscosity isotherms have yet been successful.

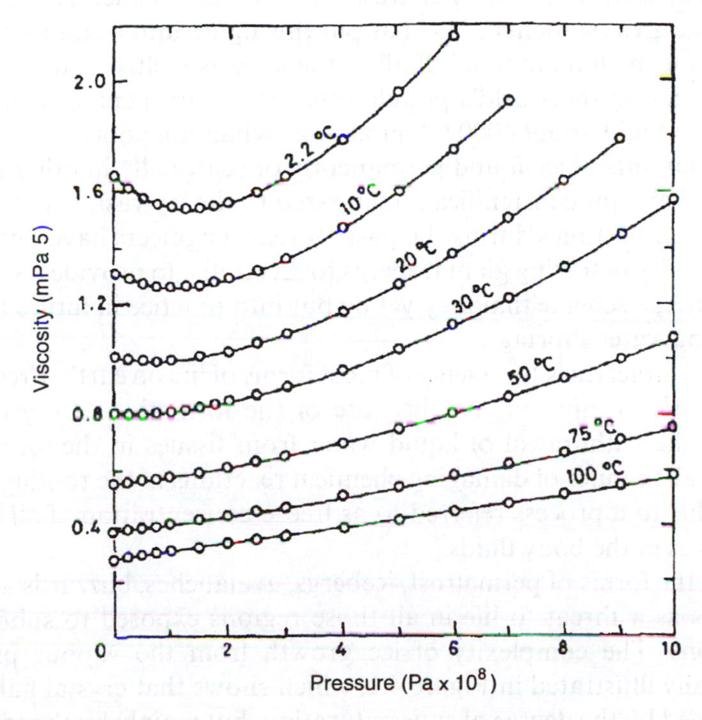

Figure 3.6 *Viscosity of water under pressure*

Chapter 4

Crystalline Water

Occurrence in the Ecosphere and Beyond

Water is the only chemical substance which is present in all three states of matter in the ecosphere. While most liquid water occurs in the oceans (10^{21} kg), the Antarctic continent alone contains 10^{19} kg of ice and snow which makes up 99.997% of all fresh water on this planet. Hiroshi Suga, one of the great students of ice, has put this figure into a comprehensible context for the human mind: If all of the ice was melted and distributed equally among the world's population, then every person would have enough to build about 4000 Olympic size swimming pools.

The amounts of ice found permanently or seasonally in other parts of the world are quite insignificant compared to the icy vastness of Antarctica. At various times during the past 50 years engineers have toyed with the feasibility of towing giant icebergs to Australia, to provide a supply of fresh water, a scheme that may yet be put into practice in future times of increasing water shortage.

By and large, ice is the enemy of most forms of life on earth. Freezing of tissue water is injurious not because of the formation of crystals but because the withdrawal of liquid water from tissues in the form of ice leads to all manner of damaging chemical reactions in the residual liquid phase, due to a process referred to as freeze concentration of all soluble substances in the body fluids.

Ice in the forms of permafrost, icebergs, avalanches, blizzards and hail also acts as a threat to life in all those regions exposed to subfreezing conditions. The complexity of ice growth from the vapour phase is graphically illustrated in Figure 4.1, which shows that crystal habits are determined by the degree of supersaturation, but mainly by the tempera-

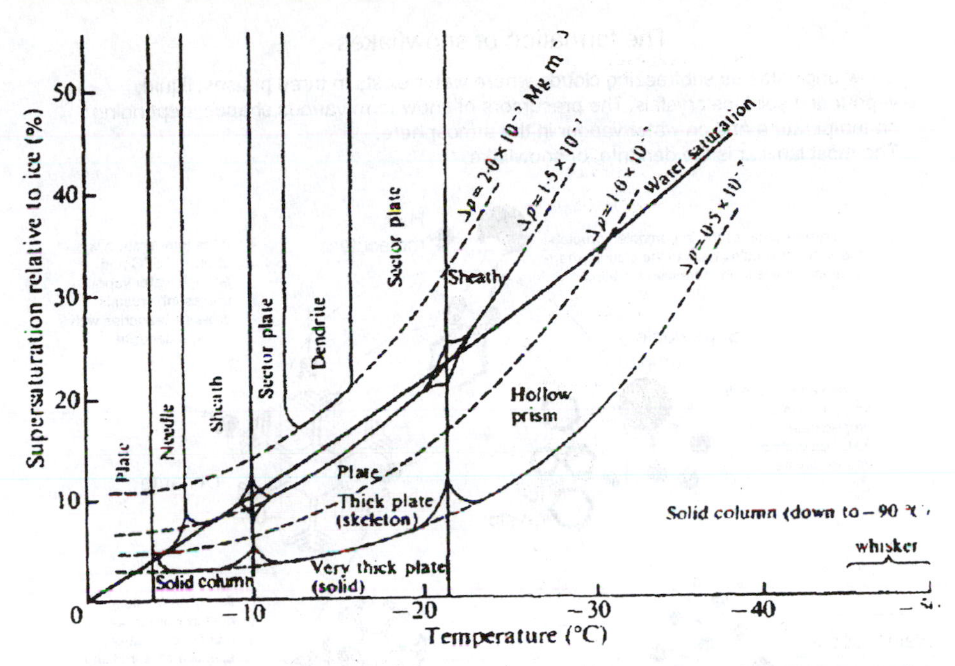

Figure 4.1 *Growth habits of ice crystals as a function of temperature and supersaturation* (Reproduced by permission of The Royal Society from Hallett and Mason, *Proc. Roy. Soc. Lond.*, 1958, **A247**, 440)

ture. The meteorological implications are illustrated in Figure 4.2 for conditions that promote the formation of snow.

Observations and studies of 'ice from the sky' go back in history to Chinese writings of the 2nd Century BC, where the number *six* became associated with water and ice. The beauty, symmetry and diversity of solid water have excited the minds of philosophers and artists throughout several centuries and are likely to continue to do so for a long time yet. Surprisingly, the formation of ice in the atmosphere has been a subject of scientific investigation only since the first comments by Kepler on its hexagonal nature.

Ice growth from the liquid phase provides even more complexities, partly because existing theories of ice nucleation cannot easily be applied to an undercooled liquid which exists in the form of an infinite hydrogen-bonded molecular network. The extensive morphological diversity of ice crystals depends on the degree of undercooling; it ranges from needles to dendrite sheets and pyramids. The disciplines of glaciology and ice engineering depend largely on what is known about the structural and mechanical behaviour of 'solid water' and its mechanisms of formation.

The existence of ice in interstellar space is now also well established. There is mounting evidence that comets are composed primarily of ice,

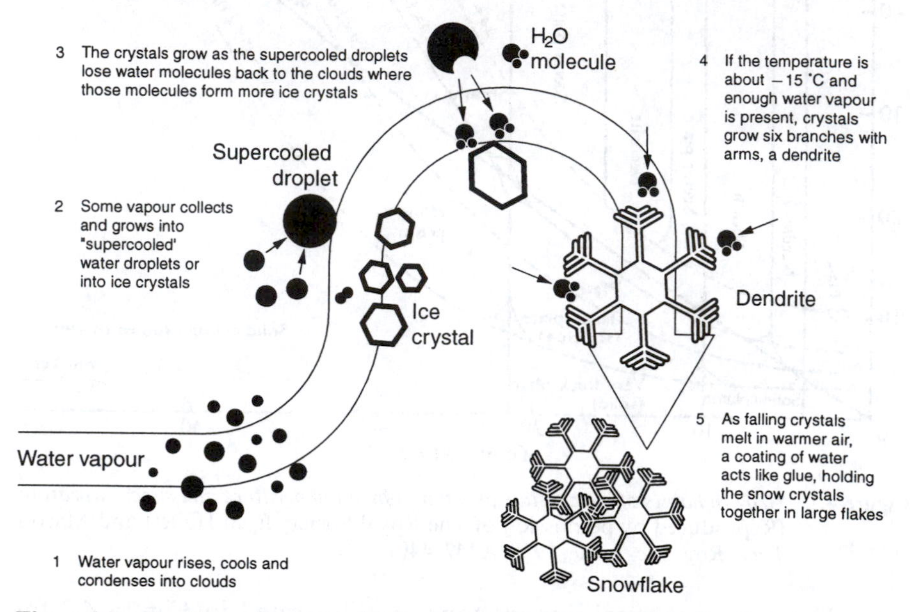

Figure 4.2 *Conditions favouring the formation of snow in the upper atmosphere* (Adapted from *USA Today*, 1999)

and that they also contain other, ice-related structures, possibly clathrate hydrates (see below). Space research has also revealed the presence of ice in the outer solar system. Thus, the stability of ice spheres of 1 km radius and of a comparable measured reflectivity has been estimated (in years) as Venus 2.2×10^2, Jupiter 2.4×10^{13} and Saturn 6.7×10^{21}.

The above brief survey of ice in our ecosphere, and far beyond, should serve to convince the reader of the importance of ice studies as valid scientific and technological disciplines. The following sections of this chapter concentrate on physico-chemical aspects of ice, while other chapters touch upon some of the ice issues that have a profound impact on the life sciences. Problems that relate to space exploration, meteorology, permafrost and ice engineering are well beyond the scope of this book.

Structure and Polymorphism

In order to learn about the molecular structure of a substance, the examination of the crystalline solid state seems to offer the most direct approach. Such an examination of ice immediately reveals the complexity of the H_2O substance. The four-point-charge model suggests a tetrahedral molecular packing with each molecule hydrogen-bonded to four nearest neighbours, and this is indeed confirmed by the results from X-ray and neutron diffraction studies. Such a regular, 4-coordinated structure, depicted in Figure 4.3 contains a large amount of empty space compared to hexagonal close-packed structures in which each molecule has 12 nearest neighbours. One important consequence is that the structure of ice is sensitive to pressure and that, apart from the 'normal' low pressure form (ice I), there exist other crystalline modifications, each stable over a limited range of pressure and temperature. The phase diagram of ice, shown in Figure 4.4 is therefore extremely complex, with some of its features still somewhat uncertain. It indicates that ice II, VIII and IX cannot be obtained directly from liquid water. It also shows that at high pressures liquid water can be made to freeze at temperatures of 80 °C, and even higher, with the crystallisation of ice VII.

All the various polymorphic forms have one feature in common with the structure in Figure 4.3: each oxygen atom is hydrogen-bonded to four other oxygen atoms. It seems that pressure alone can only distort, but not break hydrogen bonds. The differences in the various structures arise from an increasing degree of hydrogen-bond deformation with increasing pressure. The net result is a lengthening of the O–O distances and a decrease in the next-nearest neighbour distances from the characteristic value of 0.45 nm in the regular tetrahedron. For example, in ice V the hydrogen-bond donor angles range from 84° to 135° (compared with 109° 28′ in the regular tetrahedron), giving rise to a distribution of non-bonded

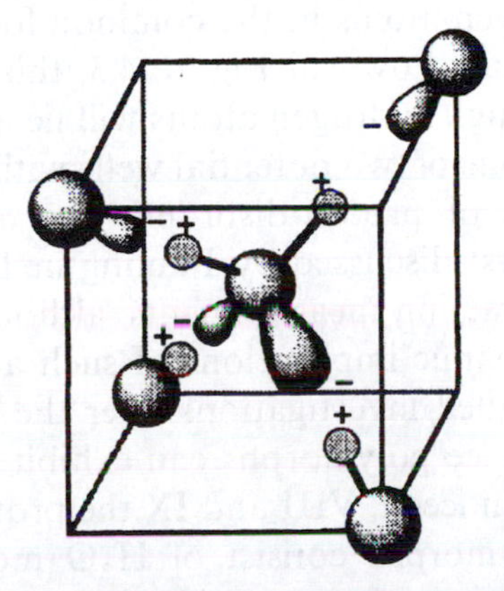

Figure 4.3 *Local structure of a water molecule surrounded by four tetrahedrally arranged neighbours, characteristic of 'ordinary' hexagonal ice*
(Reprinted from *Thermochim. Acta*, **300**, H. Suga, 118. © 1997. With permission from Elsevier Science)

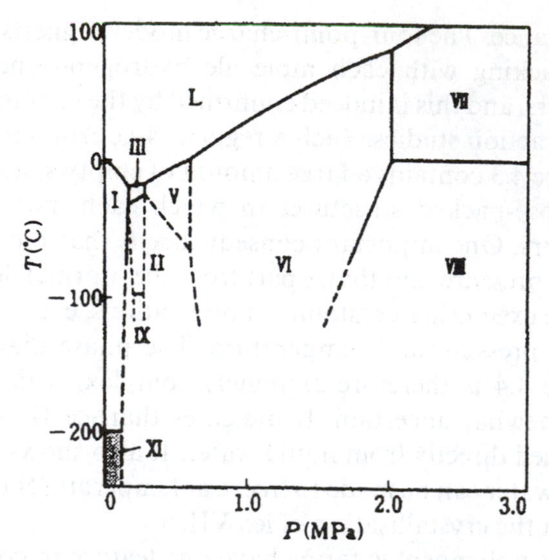

Figure 4.4 *Solid–liquid phase region diagram of H_2O ice*

(next-nearest neighbours) O–O distances ranging from 0.328 to 0.346 nm. In the high-pressure limit (ice VII and VIII) each oxygen atom is surrounded by *eight* nearest neighbours at 0.286 nm (four nearest neighbours at 0.275 nm in ice-I), of which four are hydrogen-bonded neighbours and the other four arise from the pressure-induced distortion of the tetrahedral structure. If it were not for the lengthening of the hydrogen bonds, ice-VII would be twice as dense as ice-I.

There is an additional, and major, structural complexity to be considered. While the oxygen atoms in the common form, ice-I, do indeed form a regular lattice, as shown in Figure 4.3, this is not true for the hydrogen atoms. Although hydrogen atoms will lie along the O–O axes, they can be located in one of two potential wells with an equal probability, leading to a degree of 'proton disorder'. The well-studied residual entropy of $R \ln(3/2)$, first discussed by Pauling in 1935, was later confirmed by neutron diffraction measurements, although at that time no information on the dynamic implications of such a disorder was presented (see below). Detailed investigations over the past 30 years have shown that the various ice polymorphs can exhibit differing degrees of proton ordering. Thus in ice-II, VIII and IX the protons are completely ordered, i.e. these polymorphs consist of H_2O molecules, hydrogen-bonded in a regular lattice with zero residual entropy. Presumably any observed proton disorder arises from kinetic barriers which retard the ordering processes which would otherwise lead to perfect crystals, in conformity with the Third Law of Thermodynamics. This has indeed been confirmed by the introduction of impurities which act as catalysts to the ordering process, i.e. proton jumps along the O–O axes.

Experimental studies of disorder–order transitions are bedevilled by the phenomenon of vitrification, associated with a glass transition temperature at which all dynamic motions decrease by many orders of magnitude and where disorder can then become frozen-in, at least on any practical time-scale. Practical implications of glass transitions in aqueous systems are discussed in more detail in Chapters 12 and 13. Experiments with KOH-doped hexagonal D_2O ice at low temperature and high pressure have, for instance, revealed a thermal transition at 72 K which might be either a first-order transition between ice-I and an as yet undiscovered polymorph, and/or a proton-ordering in ice-I. The 'new' modification has been designated as ice-XI. Similarly, a proton-ordering process has been observed for ice-VI, giving rise to ice-X (not shown in Figure 4.3). Recently a new modification, designated ice-XII, has been identified at pressures as high as 300 GPa. The phase diagram of ice is thus very rich, but some uncertainties remain concerning structural details.

Yet another structural modification, one which has long been known, is cubic ice, designated as ice-Ic. It is metastable with respect to 'ordinary' hexagonal ice-Ih at atmospheric pressures and it can appear in the upper atmosphere. Its hydrogen-bonded structure is of the diamond type (cf. the graphite type of ice-Ih), with the same O–O distance as ice Ih. Ice-Ic can be prepared by a variety of methods: water vapour deposition onto a cold substrate; rapid quenching of aqueous aerosol on a cryoplate; depressurisation of any high-pressure modifications of ice at 77 K, followed by warming to $\sim$160 K. All the cubic ice thus produced exhibits an irreversible transition to ice-Ih on heating at an extremely low rate in the temperature range 170–220 K; the transition is accompanied by heat evolution of $\sim$40 J mol^{-1}. The depressurisation method is the most suitable way to prepare ice-Ic in large quantities.

Investigations into ice morphology are thus likely to continue as a fertile field for researchers. Ice-IV was for many years believed to be a metastable form but has recently received more detailed study which suggests that it might be a stable phase. There is also new work under way on the effects of compression of ice-II and VI, and the conditions for proton ordering of ice-III and V have been put on a firmer basis.

To conclude this discussion of an extremely complex subject it must be emphasised that in many cases it is still difficult to differentiate between truly amorphous phases and crystalline phases with a high degree of disorder.

Ice Dynamics

From a purely practical viewpoint, the dynamic and transport properties of ice may be even more important than knowledge of its molecular structure. Details of the ice lattice dynamics are revealed in the infrared and Raman spectra which appear extremely complex. The high dielectric permittivity of ice-I (~ 100) is a measure of the reorientation of molecular dipoles in an applied electric field. As the frequency of the field is increased to above 10 kHz, the dielectric permittivity drops sharply to 3.2. This relaxation frequency is a measure of the frequency of the molecular rotation. Since an H_2O molecule in the ice lattice can only rotate by virtue of the presence in the lattice of crystal defects, this relaxation frequency depends sensitively on the degree of purity of the sample. Another property that is sensitive to impurities is the electrical conductance, which increases exponentially with temperature. The mechanism of conductance therefore resembles that of a semi-conductor rather than that of an ionic conductor.

The peculiar electrical properties of ice have led to detailed studies of the origins of ionic and molecular motions in the crystal. Two types of crystal defects can arise in ice; they are shown in Figure 4.5. A pair of ionic defects ($H_3O^+ + OH^-$) is formed when a hydrogen atom migrates from one oxygen atom to its neighbour along the hydrogen bond axis. These two defects will become separated by subsequent proton jumps. The other type of defect arises through the rotation of a water molecule

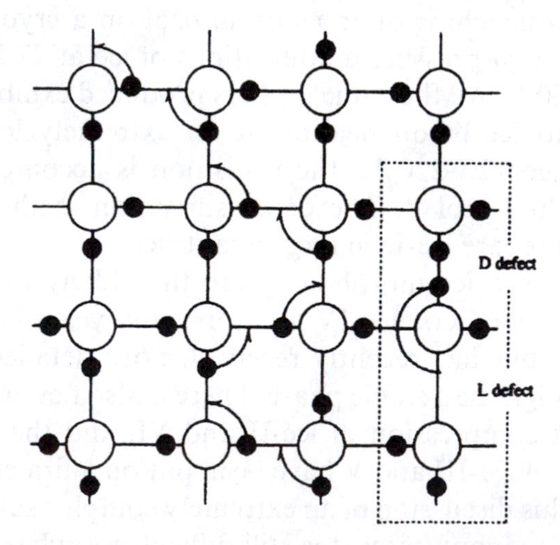

Figure 4.5 *Proton rearrangements in a two-dimensional representation of an ice lattice, showing the generation of D and L lattice defects*
(Reprinted from *Thermochim. Acta*, **300**, H. Suga, 122. © 1997. With permission from Elsevier Science)

through $2\pi/3$; the actual movement can be through an oblique proton jump. This movement gives rise to a pair of orientational defects which will become separated by further oblique proton jumps. This type of defect, being nonionic, cannot contribute to the electrical conductivity, but it does contribute to the dielectric permittivity and to self-diffusion.

The above properties of ice, and others, are of great importance in several scientific disciplines and also in technological, ecological and engineering applications. Thus, ice reveals a complex elastic behaviour with some of the features of a plastic substance, a fact of considerable importance in the behaviour of glaciers and soils in permafrost regions. Other important properties include the phenomena of pressure melting, crystal nucleation and growth, especially in the upper atmosphere, and effects of ageing on surface structure. Although detailed descriptions of these properties are quite beyond the scope of this book, the subject of nucleation and growth of ice crystals from the liquid phase receives attention in several chapters.

Clathrate Hydrates

A chapter on crystalline water would be incomplete without some mention of a group of 'ice cousins', the clathrate hydrates, also known as gas hydrates. Like the ice polymorphs, they are crystalline solids, formed by water molecules, but hydrogen-bonded in such a way that polyhedral cavities of different sizes are created which are capable of accommodating certain types of 'guest' molecules.

Structurally the clathrates differ from ice only in the configuration of neighbouring water molecules. Thus, in ice neighbour H_2O molecules are in a *gauche* configuration, whereas in the clathrate hydrates they are in a

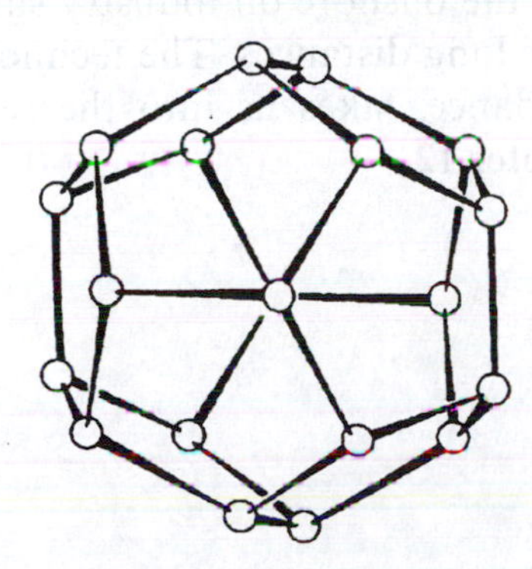

Figure 4.6 *Pentagonal dodecahedral clathrate cage, as it appears in Type-I clathrate hydrates. Circles denote oxygen atoms*

cis configuration; Figure 4.6 shows a typical clathrate arrangement of water molecules. The *cis* configuration ensures that neither the donor nor the acceptor orbitals of the H_2O molecule are directed towards the centre of the cavity; hydrogen bonds can only form along edges of the cavities. Given that the hydrogen-bond lengths and energies closely resemble those characteristic of ice, the important result is that the water molecules in clathrates possess fewer configurational degrees of freedom than they do in ice.

The simplest cavity which can be formed by a set of tetrahedrally oriented water molecules is a pentagonal dodecahedron, but such structures cannot fill space completely. The clathrate hydrates are therefore composed of combinations of different types of cavities, bounded mainly by pentagonal and hexagonal faces. Their formation necessitates distortions in the ideal tetrahedral hydrogen-bond angles. A commonly found structure is the so-called type-II hydrate, in which the cubic unit cell is composed of 16 pentagonal dodecahedral and 8 hexakaidecahedral cages formed by 136 water molecules. Most of the guest molecules in the type-II hydrates are contained in the larger cages, so that the stoichiometric formula for the full occupation by a guest molecule G is $G.17H_2O$.

The empty clathrate lattice structure is unstable relative to ice but is stabilised by van der Waals interactions with the guest molecules. There are lower (0.4 nm) and upper (0.75 nm) size limits for guest molecules, as well as certain other chemical constraints. The most studied hydrates are those of the noble gases, lower hydrocarbons, H_2S, CO_2, N_2, O_2, Cl_2, ethylene oxide, trimethylene oxide, acetone and tetrahydrofuran.

Although the phenomenon of clathrate formation has been applied in practice, e.g. for the removal of residual water from sulfuric acid by pumping in propane gas and cooling, a much more important aspect is the prevention of clathrate crystallisation in natural-gas pipelines, a problem that has dogged the offshore oil industry since wet natural gas was first transported over long distances. The technology of crystallisation prevention, or avoidance, takes us into the realm of metastable water, the subject of Chapter 12.

Chapter 5

The Structure of Liquid Water

Diffraction Methods

The very term 'structure of a liquid' requires some explanation and justification. Generally a structure is defined by the spatial coordinates of atoms or molecules and contains an element of periodicity. A liquid, on the other hand, is characterised by the random diffusion (Brownian motion) of molecules and by the absence of periodic order. In other words, the liquid state cannot be described by a set of molecular coordinates.

Diffraction techniques that had led to many important advances in the elucidation of crystal structures were later also applied to liquids and it was found that liquids, like crystalline solids, scatter radiation, and that fairly well defined patterns of electron (or neutron) intensity could be obtained. The processing of such data is of necessity different and more complex, because of the motions of the molecules while the measurement is being made. Thus, what in the solid state is simply measured as an electron density, in the liquid state becomes a *probability electron density*, indicating the level of probability that certain positions are occupied by atoms at certain times, but without specifying the lifetime of the atom at the measured location.

The diffraction experiment consists of subjecting a thin sample of liquid to radiation of a given wavelength λ, where λ is of the order of the molecular spacings, and determining the intensity of the scattered radiation $I(\theta)$ as a function of the scattering angle θ. This quantity is the Fourier transform of one of the most basic properties of a liquid, the radial distribution function $g_{ij}(r)$, already discussed in Chapter 2, which measures the probability of finding an atom j at a distance r from a given atom i. Thus, if $g_{ij}(r)$ is large for a given value of r, there is a preferred atomic spacing, rather than a completely random arrangement of atoms, for which case $g_{ij}(r) = 1$. While the first X-ray diffraction studies on liquid water date from the 1930s, recent developments of neutron sources and instrumentation have added a new dimension to the structural information, which can be obtained from diffraction experiments. Whereas electrons are scattered mainly by oxygen atoms, neutrons are also scattered strongly by hydrogen. An additional useful feature is that the atomic neutron scattering power is isotope specific. The technique of 'isotope substitution' has been particularly useful in studies of 'structure' in water and aqueous solutions, as it enables certain interactions to be made selectively 'invisible', thus facilitating data processing.

The additional information that is potentially obtainable from neutron scattering data brings with it severe interpretation problems. Thus, X-ray results provide a picture of the spatial disposition of oxygen atoms, through $g_{OO}(r)$. The probable position of the hydrogen atoms can only be inferred from what is known about the O–H valence bond, the H–O–H valence angle and assumed details of the hydrogen bond. The neutron data, on the other hand, can provide the additional pair distribution functions $g_{OH}(r)$ and $g_{HH}(r)$. The three distribution functions, taken together, provide the orientational information which permits the complete three-dimensional structure to be elucidated, i.e. $g_{ij}(r,\Omega)$.

Theoretical Approaches

The distribution of molecules in a fluid is governed by the energy of interaction between them. The nature of the interaction, in turn, sensitively depends on the molecular geometry and charge distribution. In the case of the water dimer, the hydrogen bond is the predominating contributor to the interaction energy. Also, from the tetrahedral disposition of the electron density in the water molecule one could make a well-informed guess as to the molecular structure of ice. Further, since the latent heat of fusion of ice is quite small, one could guess further that in liquid water some of the tetrahedral dispositions of molecules might persist, and that the spacings between oxygen atoms should be dominated by the hydrogen bond length, i.e. of the order of 0.3 nm.

One of the major shortcomings of present-day theories of the liquid state is the necessity of expressing the total interaction energy between the molecules by the sum of pairwise molecular interactions, such as $u_{ij}(r)^*$,[†] when it is known that the characteristic behaviour of a liquid is in fact due to simultaneous interactions between *many* molecules. However, the mathematical problems of coping with triplet and higher order interac-

[†] From here on, the subscript $_{ij}$ will be omitted in the text, unless it is specifically required for clarity.

tions are so severe, that such systems have to be treated in terms of a summation of three (or more) pairwise interactions.

There exist formal relationships between $u(r)$ and $g(r)$. Although the former is the more basic quantity, it is the latter which is more easily accessible by experiment and which provides the structural information. It is more common to calculate $g(r)$ from X-ray or neutron scattering experiments and to work backwards and derive a pair potential function that is consistent with the radial distribution function. In practice, therefore, $g(r)$ is the most informative feature of a liquid. This is well illustrated in Figure 2.4 which compares the radial distribution functions of liquid water and liquid argon, derived from X-ray diffraction experiments. The abscissa has been scaled so that a very direct comparison becomes possible: $R^* = R/\sigma$, σ being the van der Waals diameter of the molecule (0.28 nm for water and 0.34 nm for argon). The sharp peak near $R^* = 1$ indicates that in the liquid state there is a preferred distance of separation which is very close to that observed in the crystal ($R^* \equiv 1$). The number of such nearest neighbours of any given molecule is called the coordination number and is given by

$$n(r') = \rho \int_0^{r'} g(r)\, 4\pi r^2 \mathrm{d}r$$

where r' can be taken to be the position of the first minimum in $g(r)$, and ρ is the density. By performing the integration for the two curves in Figure 2.4, one obtains $n \cong 10$ for argon and $n \sim 4.4$ for water. Furthermore, from scattering experiments at increasing temperatures it becomes apparent that for argon n decreases fairly rapidly, whereas for water n increases slowly, to reach 5 at the boiling point. Another distinguishing feature in the two $g(R^*)$ curves is the position of the second peak. For argon this occurs close to $2R^*$, whereas for water it is found near $1.6R^*$.

We can now attempt to interpret these findings in terms of the remnants of structure that persist in water and in argon but, before doing so, let us note that any such structure must be of a relatively short range, because the oscillations in $g(r)$ die away after the third peak. This means that any given molecule can 'see' order only as far as three molecular diameters. Beyond this the molecular arrangement becomes uniform and random, with no preferential intermolecular spacings or orientations.

Argon, in the crystalline state, is close packed with a coordination number of 12. On melting, this arrangement is slightly perturbed, but there remain relatively high concentrations of atoms at about σ, 2σ and 3σ. The coordination numbers are slightly smaller than those in the crystal, because of the expansion on melting. In water, however, the electronic distribution dictates the *tetrahedral* (instead of hexagonal close-packed) arrangement of molecules round a given central molecule. Thus, although the first neighbour of any molecule will be found at a distance σ, the next-nearest neighbour of any molecule will be found at a distance σ, the next-nearest neighbour will be at a distance 2σ sin $(\frac{1}{2} \times 109° 28') \sim 0.45$ nm, or $R^* = 1.6$. With its 4-coordinated structure, the ice lattice contains a large amount of empty space (see Figure 4.2), and upon melting there is a partial collapse of this very open structure, reflected in the observed *increase* in density. This collapse continues as the temperature is raised, and hence n increases slightly, despite the normal thermal expansion of the liquid. If the scattering experiments were to be performed at constant density of the liquid, the observed increase in n would be much more marked.

One further detail of $g(r)$ is of importance, namely the widths of the peaks. They are the real indicators of 'structure' in the liquid. Thus, a narrow peak indicates little perturbation about the mean value of R^*, as defined by the peak maximum. Judged by this criterion, water is more structured than argon, because the first peak (at half maximum height) is much narrower than the argon peak. Similarly, the depths of the minima between peaks are indicative of structure; $g(r) = 0$ means that at that value of r there is a zero probability of finding an atom. This result is also illustrated in Figure 2.4, where it is seen that for $R^* < 0.9$, $g(R^*) = 0$, because the molecules are almost incompressible and they strongly repel each other as they approach a distance of separation given by σ.

Let us now consider the additional experimental information that can be derived from neutron scattering. The three atomic pair distribution functions are shown in Figure 5.1(a–c). As expected, the shape of $g_{OO}(r)$ closely resembles that obtained from X-ray experiments. The oxygen–hydrogen $g_{OH}(r)$ in Figure 5.1(b) has a strong, narrow peak close to 0.1 nm, characteristic of the O–H valence bond length. The peak area shows two H atoms at this distance, consistent with the formula H_2O. The second peak, at 0.185 nm is characteristic of the nearest hydrogen-bonded, i.e. intermolecular O–H neighbour. The third peak, centred on 0.325 nm, is due to non-hydrogen-bonded neighbours.

A study of $g_{HH}(r)$, in Figure 5.1(c), completes the picture. Once again, the first sharp peak arises from the intramolecular H–H separation, given by the molecular geometry. The second peak, at ca. 0.24 nm corresponds to the closest H–H distances between two hydrogen-bonded molecules. More distant peaks report on H–H distances in

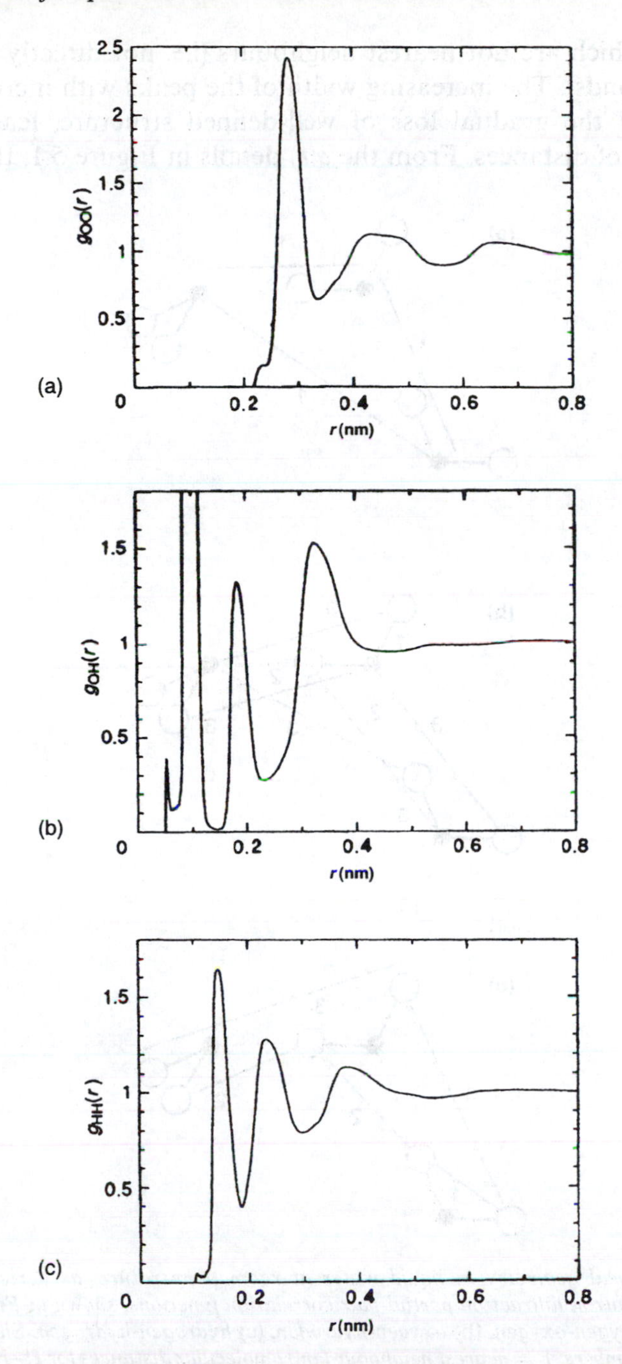

Figure 5.1 *Partial pair correlation function* $g_{ij}(r)$ *for water at room temperature, as indicated on the ordinate axes. For details, see text*
(J.L. Finney and A.K. Soper, *Chem. Soc. Rev.*, 1994, **23**, 4. Reproduced by permission of the Royal Society of Chemistry)

molecules which are not nearest neighbours (i.e. not directly linked by hydrogen bonds). The increasing width of the peaks with increasing r is indicative of the gradual loss of well-defined structure, leading to a distribution of distances. From the $g(r)$ details in Figure 5.1, the mutual

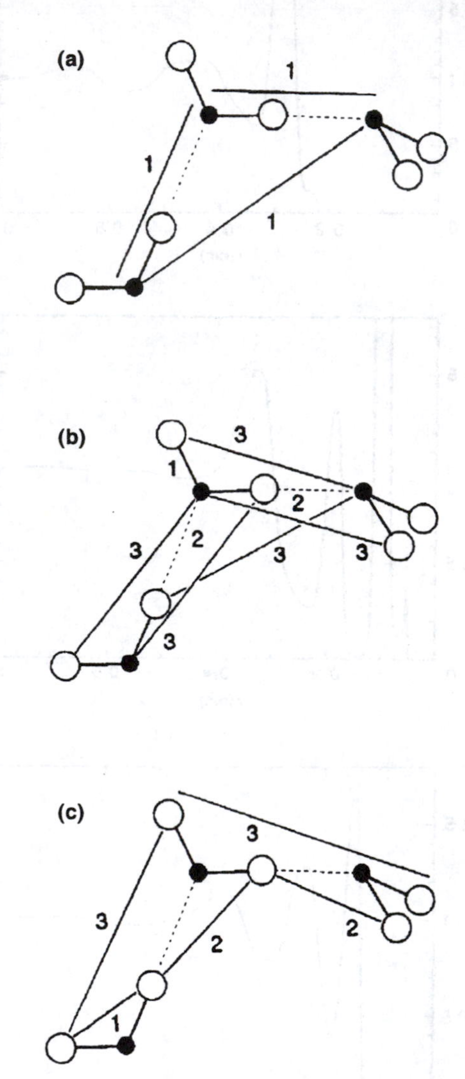

Figure 5.2 *Local geometry in liquid water at room temperature, as derived from the neutron diffraction partial pair correlation functions, shown in Figure 5.1:* (a) *oxygen-oxygen,* (b) *oxygen-hydrogen,* (c) *hydrogen-hydrogen. Significance of numbers: 1 = nearest neighbour (intramolecular distances for H–H and O–H), 2 = next-nearest intermolecular neighbour, 3 = next-distant intermolecular neighbour (corresponds to third peak in* g(r)*)*
(J.L. Finney and A.K. Soper, *Chem. Soc. Rev.*, 1994, **23**, 5. Reproduced by permission of the Royal Society of Chemistry)

orientations of neighbouring water molecules can be calculated; they are shown for a trimer in Figure 5.2.

The Concept of Water-structure Making and Breaking – Computer Simulation

The literature on aqueous solutions contains abundant references to water-structure making and breaking by dissolved species. Let it be remembered that 'structure' is described by distances and angles *only*. Taking the $g(r, \Omega)$ functions as examples, symptoms of structure-making should be a narrowing of the peaks, with possible shifts of some peak positions. A decreasing temperature would produce the most obvious structure-making effect. This is indeed found to be the case for $g_{OO}(r)$, as measured by X-ray diffraction. Neutron data are not yet extensive enough to permit definitive interpretations of temperature effects. We shall be examining the evidence for and against structure-making and breaking speculations in subsequent chapters.

Reference has already been made to relationships between $g(r)$ and $u(r)$. In the limit of the ideal gas, there are no interactions, so that $u(r) = 0$. For a dilute gas, and also for a dilute solution, $g(r) = \exp[-u(r)/kT]$ which, in the limit of $r \to \infty$, approaches unity (see Figure 2.4). For a more concentrated fluid, $g(r)$ can be expressed in terms of a virial expansion in the density:

$$g(r) = (\exp[-u(r)/kT])\{1 + B\rho(r) + C\rho^2(r) + \ldots\} \tag{5.1}$$

where $B, C \ldots$ are the virial coefficients.

Fast computers have assisted in the development of new techniques for probing interactions and structures in liquids; they are referred to as simulation methods. First a 'reasonable' pair potential function $u(r)$ is chosen to describe the interaction of molecules which are placed randomly in a given volume. There are then two ways of proceeding:

(1) Molecules are displaced by small amounts dr and the energy of the system is calculated from the sum of $u(r)$ values, taken over all pairs of molecules. The procedure is repeated, accepting only those displacements that lead to a lowering of the total energy. The equilibrium state is assumed to be the one of lowest energy, and $g(r)$ for that state can be calculated from the corresponding coordinates of the molecules. This procedure is called the Monte Carlo technique and provides information about equilibrium properties only.

(2) In the molecular dynamics (MD) method, energy is supplied to an assembly of molecules, which are allowed to move, and the system

is allowed to find its own equilibrium state. The coordinates are sampled at fixed intervals of time (characteristically 10^{-15} s), so that the time evolution of equilibrium can be studied. In effect, the method relies on the solution of the equations of motion, given a model interaction energy $u(r)$. Apart from giving the equilibrium thermodynamic properties of the system, the MD simulations also yield time-dependent properties of the system, such as the self-diffusion coefficient.

Both the above simulation techniques have been applied to investigations of water and aqueous solutions, with variable results. Their advantages include the possibility of exploring regions of phase space, e.g. very high-density fluids, which are not readily accessible to experiment, and they can also provide graphic descriptions of molecular distributions and motions in complex systems. A disadvantage is that, with present computing facilities, the number of molecules in an assembly is limited; typically arrays of $8^3 = 512$ water molecules have been studied. While this may just be sufficient to study most of the properties of a one-component liquid, where long-range effects (such as dipole-induced orientations) are non-existent or negligible, it is not necessarily the case for aqueous solutions. A problem still of topical interest involves the interactions between nonpolar molecules in aqueous solution, especially those where long-range effects are suspected (see Chapter 6). The simplest way of studying such a system by computer methods is through a consideration of two solute particles (e.g. neon atoms) in $(512-2)$ molecules of water, corresponding to a solute concentration of approximately 0.2 mol kg^{-1} which is physically quite unrealistic in view of the very limited solubility of the noble gases. The most serious limitation of all computer methods is that the quality of the output can be no better than the quality of the interaction potential $u(r)$ used in the calculations.

Several methods have been applied to the estimation of 'reasonable' pair potential functions. Some are phenomenological techniques, which use empirical information and adjustable parameters to obtain $u(r)$ or $u(r,\Omega)$. The following example is a popular approach: R is a given coordinate set for which the corresponding energy $u(R)$ is given by the additive sum of five contributions:

$$U(R) = \tfrac{1}{2}\sum_{\substack{\text{bond}\\ \text{lengths}}} K_b(b-b_0)^2 + \tfrac{1}{2}\sum_{\substack{\text{bond}\\ \text{angles}}} K_\theta(\theta-\theta_0)^2 + \tfrac{1}{2}\sum_{\substack{\text{torsional}\\ \text{angles}}} K_\phi[1+\cos(n\phi-\phi_0)]$$
$$+ \sum_{\substack{\text{non-bonded}\\ \text{pairs, } r<0.8\text{ nm}}} [A/r^{12} - B/r^6 + q_1q_2/4\pi\varepsilon_0] + \sum_{\substack{\text{hydrogen}\\ \text{bonds}}} (A'/r^m - C'/r^p)f(\theta',{}'') \quad (5.2)$$

The first term describes the interatomic bond lengths b, referred to their equilibrium values b_o, based on the harmonic oscillator model. The second term describes the bond angle librations about their mean values θ_0, with K_θ as the force constant. The third term describes the torsional angle potential, based on the assumption that rotation about valence bonds is not free, but subject to steric barriers; n is the multiplicity (degeneracy) of the torsional angle; e.g. $n = 3$ for the C–C bond in ethane.

The fourth term in equation (5.2) accounts for short-range interactions between non-bonded atoms in terms of the familiar Lennard–Jones 6–12 potential and a Coulombic term to allow for electrostatic interactions between atomic charges separated by a distance r_0 and placed in a medium of dielectric permittivity ε. Non-bonded interactions can also be described by expressions other than the popular 6–12 potential. For instance, an exponential three-parameter form (Buckingham potential)

$$\mathrm{u}(r) = A \exp(ar) - Cr^{-n}$$

has a sounder theoretical basis than the Lennard–Jones expression.

The description of a realistic hydrogen bond potential introduces problems. It is expressed in equation (5.2) as a van der Waals-type expression, modified by some function of the donor and acceptor group valence angles.

Other methods include (i) strictly *ab initio* approaches in which no adjustable parameters are required, (ii) semi-empirical quantum mechanical methods which require knowledge of various parameters, including details of the molecular geometry.

General Significance of $u(r)$ and $g(r)$

Both $u(r)$ and $g(r)$ are of intrinsic interest because they describe intermolecular interactions and distributions (structure), but the two properties are also of more general importance, because they form the bridge between structural measurements, such as scattering of radiation, and equilibrium thermodynamic functions. Thus, all the latter can be expressed in terms of various derivatives of $u(r)$ and $g(r)$. For instance, the pressure of a gas (or the internal pressure of a liquid) is given by

$$P = kT\rho - \frac{\rho^2}{6}\int_0^\infty r\frac{\mathrm{d}u(r)}{\mathrm{d}r}g(r)4\pi r^2\,\mathrm{d}r \qquad (5.3)$$

where the first term on the right-hand side describes ideal gas behaviour, and the second term accounts for the interactions between molecules. Similarly, the heat capacity at constant volume, C_V, is given by

$$C_V = \frac{3}{2}k + \frac{\rho^2}{2}\int_0^\infty \left[\frac{\partial g(r)}{\partial T}\right]_V u(r)4\pi r^2\,\mathrm{d}r \tag{5.4}$$

and the isothermal compressibility κ_T by

$$\kappa_T = \frac{1}{\rho kT} + \frac{1}{kT}\int_0^\infty [g(r) - 1]4\pi r^2\,\mathrm{d}r \tag{5.5}$$

Equation (5.5) is of particular value, because it does not require knowledge of $u(r)$ and can therefore be used, in conjunction with equation 5.3 as a test for the quality of a given interaction potential.

The preceding discussion of the physical properties of water and of its place among liquids has illustrated the complexity of the intermolecular interactions that are responsible for the very atypical behaviour observed. It has so far been impossible to reconcile $u(r)$ and $g(r)$ with *all* the known physical properties and to arrive at a molecular description of water which will account correctly for its behaviour over the whole liquid range from the undercooled state to the critical point and beyond. While it is beyond the scope of this book to consider the properties in detail, we will review a few of the seeming paradoxes which make the formulation of a credible molecular description so difficult.

It is often said that liquid water is ice-like, and the low heat of fusion of ice and the X-ray radial distribution function, shown in Figure 2.4, lend support to this claim. On the other hand, if the liquid possesses structural features that closely resemble those of the crystalline solid, then why is water so easily undercooled? Provided that adequate precautions are taken to prevent ice nucleation by particulate impurities, liquid water can be cooled in the laboratory to $-39.5\,°\mathrm{C}$, and to much lower temperatures under conditions of applied pressure. The physical properties of undercooled water are even more bizarre than those of the liquid at ordinary temperatures; they will be reviewed in Chapter 12.

The results in Table 3.2 suggest that the transport properties of water seem to conform quite well to what one might expect for a liquid. Thus, the coefficients of viscosity and diffusion are not in any way abnormal. However, the numbers in Table 3.2 also hide quite remarkable pressure and temperature effects. Bearing in mind the hydrogen-bonded nature of water, it is in any case remarkable that the experimental results can be accounted for on the basis of molecular transport mechanisms that should not really apply to such a liquid.

Another unsolved problem relates to the so-called vapour pressure isotope effect. It is often stated that the deuterium bond is stronger than the hydrogen bond, hence D_2O has higher melting and boiling points than H_2O. This is explained in terms of the lower zero-point energy of D_2O. What is not so easy to explain is that just above 200 °C the vapour pressure curves of the two liquids cross over, so that at high temperatures D_2O is more volatile and it also has a lower critical temperature than H_2O.

The Raman and infrared spectra also pose unresolved problems: if the fundamental bands can indeed be decomposed into two bands that show the observed temperature and pressure dependence, then this constitutes a strong case in favour of a mixture model (two distinguishable molecular species) for water. On the other hand, no such model has yet been devised that can account for *all* physical properties of water. Also, the results from scattering and computer simulation experiments cannot be reconciled with mixture models.

Present 'Best Guess'

What, then, is the present consensus regarding a credible molecular description of liquid water? One can do no better than to quote Henry Frank who contributed more than anyone else to shaping such a consensus.

> *While there has been no dearth of suggestions about what water* might *be like in order to display this or that set of properties, it is only recently . . . that it has begun to be possible to draw useful inferences of what water* must *be like in this or that respect. Recent developments do, however, seem to portend real progress towards a comprehensive, self-consistent model of water structure upon which we will be justified in expecting successive approximations to converge.*

This was written in 1971. Almost three decades later there, progress has indeed been made, but it is patently clear that, despite its molecular simplicity, water in the bulk liquid still presents many major puzzles to physical and life scientists.

It can now be reasonably concluded that, on an appropriate time-scale, liquid water retains several of the *structural* characteristics of ice: it is extensively hydrogen-bonded, although the strict tetrahedrality of ice has been lost. On a picosecond time-scale it can be regarded as an infinite three-dimensional network of H_2O molecules, connected by hydrogen bonds. In its *dynamic* behaviour, water resembles most other liquids: rotational and translational correlation times are short, in the order or 1–5 ps, indicating high hydrogen bond exchange rates.

Earlier theoretical treatments were based on the assumption that, with rising temperature, hydrogen bonds become progressively ruptured, so that at its normal boiling point, no hydrogen bonds remain. Such an assumption does, however, beg the question of the nature of the interactions that are responsible for the very high critical temperature, 374 °C, quite atypical of a small molecule. A related puzzle concerns the H_2O/D_2O vapour pressure isotope effect. Most differences between the two isotopic forms are rationalised in terms of the lower zero-point energy of D_2O, but this cannot account for the crossing over of the vapour pressure curves, giving D_2O a *lower* critical temperature than H_2O.

A comparison of the diffusional properties of the two isotopic species points to yet another puzzle. If the difference in properties were due to zero-point energies or moments of inertia, the self-diffusion coefficients would not show such marked differences as have been measured.

The current consensus view of water seems to be that water can be treated as a three-dimensional hydrogen-bonded network, with the bond lengths and angles becoming increasingly distorted with rising temperature, but without a significant number of non-bonded H_2O molecules. Perhaps recent and current research into the strange behaviour of undercooled and glassy water, to be discussed in Chapter 12, will provide further revelations about this eccentric liquid. It must be borne in mind, however, that many current researches lean heavily on computer simulation methods and probe regions of P–V–T space that are not now, and may never become, accessible to experimental verification. In closing this discussion of water structure, it is worth noting that of the published $u(r)$ functions, numbering upwards of 20, none is able to account for *all* the properties of noncrystalline water over the whole of P–V–T space.

Chapter 6

Aqueous Solutions of Simple Molecules

Molecular Interactions in Solution

For the purposes of the following discussion, 'simple' molecules are defined as those that are unlikely to interact with water or with other solution species by long-range effects and which are not able to participate in O–H···O hydrogen-bonding. It will, however, quickly transpire that such 'simple' molecules do in fact present some of the most intractable problems in aqueous solution chemistry and also have important biochemical ramifications.

Formally the properties of solutions can be accounted for in terms of three contributions to the total potential energy: solvent–solvent, solvent–solute and solute–solute effects. In a dilute solution the measured properties are dominated by those of the solvent which make the experimental investigation of solvation and solute–solute interactions difficult, even in favourable cases. Solutions of electrolytes are further complicated by the fact that the solute consists of at least two distinguishable species. Most of the published information about solute–water interactions (hydration) is still based on circumstantial experimental evidence or on comparisons of water with some reference solvent, although developments of neutron scattering methods have provided a powerful tool for direct studies of hydration (see below).

Let us consider an ensemble containing N molecules of water in a given volume, and let us replace one water molecule by a solute molecule, so that the system now contains $(N-1)$ water molecules and 1 solute molecule. The difference in the interaction energies of these two systems is the contribution due to solute hydration. If we now replace another water molecule by a solute particle, i.e. the system now contains $(N-2)$ water molecules and 2 solute particles, then an additional term appears in the total energy, due to solute–solute interactions. At even higher solute concentrations triplet and multiple-interaction terms must be taken into account.

In terms of thermodynamics, therefore, the concentration dependence of a given property is a measure of solute–solute effects, whereas the same property extrapolated to the limit of infinite dilution provides a measure of solute–solvent (hydration) effects. For electrolytes the laws of electrostatics provide a guide of how such an extrapolation should be performed (the Debye–Hückel limiting law), but there are no such rules for nonelectrolyte solutions. It is difficult to estimate the concentration that will in practice correspond to infinite dilution, where the measured properties no longer exhibit a significant concentration dependence. Until not so long ago, for instance, it was believed that the solubility of hydrocarbons in water was so low that the ideal solution laws would apply up to saturation (i.e. no solute–solute interactions). However, very high-precision vapour pressure measurements on aqueous solutions of benzene have established that, even at concentrations appreciably below saturation, deviations from Henry's law occur. Of course such effects only contribute in a very minor way to the measured vapour pressure, but their existence demonstrates that even in very dilute solutions ($\sim$0.01 mol%) benzene molecules exert a net attraction on each other, and this must raise the question of the range of intermolecular forces in solutions where long-range electrostatic effects are absent.

The thermodynamic behaviour of dilute solutions is commonly approached via their colligative properties: vapour pressure, osmotic pressure, depression of freezing point, etc. For the purposes of this discussion it is sufficient to summarise the relationships between the various measured properties and the solution concentration. Thus, the chemical potential of water, μ_1, is related to the solute concentration c_2 by

$$\mu_1 - \mu_1^0 = -\frac{RTV_1^\bullet c_2}{M_2\{1 + B_2M_2c_2\}} \tag{6.1}$$

where $V_1^\bullet$ is the molar volume of pure water, M_2 is the solute molecular weight and B_2 is the so-called second virial coefficient which measures the deviation from ideal behaviour resulting from solute–solute interactions; μ_1 is the chemical potential referred to some standard state, e.g. pure water. Also, a change in the chemical potential $d\mu_1$, is given by

$$d\mu_1 = RT\, d\ln a_1 \tag{6.2}$$

in terms of the water activity a_1. For a solution which obeys the ideal solution laws a_1 can be equated to the concentration c_1. For a *very* dilute solution at its melting point $T_m^\bullet$:

$$-\ln a_1 = \frac{\Delta H_f^\bullet}{RT_m^2\theta} \tag{6.3}$$

where $\Delta H_f^\bullet$ is the latent heat of fusion of ice at its normal melting point, 273.15 K, and θ is the melting point depression. For a *real* dilute solution the right hand side of equation (6.3) is the first term of a power series in θ.

The chemical potential μ_1 can also be expressed in terms of the solution osmotic pressure π and the partial molar volume V_1:

$$\mu_1 - \mu_1{}^\circ = -\pi V_1$$

Once again, for a *very* dilute solution $\pi = RTc_2/M_2$, and for a *real* dilute solution

$$\pi = \frac{(RTc_2}{M_2} + B^*c_2^2 + \ldots) \tag{6.4}$$

where B^* is the osmotic second virial coefficient which is related, but not equal, to B in equation (6.1).

The second virial coefficients form the link between experimentally accessible thermodynamic quantities or interaction energies and the radial distribution functions, as already discussed in Chapter 5. Formally B^* expresses the contribution of solute–solute pairwise interactions to the total free energy of the solution, and

$$B^* = -\tfrac{1}{2}\int_0^\infty [g_{22}(r) - 1]4\pi r^2 \mathrm{d}r \tag{6.5}$$

Furthermore, for a dilute solution

$$g_{22}(r) = \exp\,[-W_{22}(r)/kT] \tag{6.6}$$

Here $g_{22}(r)$ measures the distribution of solute molecules in the solution. If the solute molecules are distributed randomly, then, as shown in Figure 5.1, $g_{22}(r)$ is equal to unity. One or several peaks indicate preferred 'structures', which result from solute–solute interactions, expressed by $W_{22}(r)$, the so-called potential of mean force. $W_{22}(r)$ resembles $\mu_{22}(r)$, the chief difference being that the latter refers to the interaction of two molecules *in vacuo*, whereas the former measures the interactions of the molecules in a solvent medium, the molecular properties of which have been averaged out, i.e. the solvent is treated as a continuum and not as discrete molecules. It therefore enters into $W_{22}(r)$ only by virtue of its macroscopic properties, such as the dielectric permittivity.

Classification of Solution Behaviour – Thermodynamic Excess Functions

A useful way of representing the thermodynamic behaviour of liquid mixtures is through the concept of *excess* functions. Thus the excess Gibbs free energy change of mixing of a binary mixture is defined as

$$G^{\mathrm{E}} = G_{\mathrm{expt}} - G_{\mathrm{id}} \tag{6.7}$$

where G_{expt} is the experimentally determined free energy of mixing and G_{id} is the free energy calculated on the basis of an ideal mixture:

$$G_{\mathrm{id}} = RT(x_1 \ln x_1 + x_2 \ln x_2)$$

where x_i is the mol fraction of species i. G^{E} therefore measures the contribution of interactions between components 1 and 2. Other excess thermodynamic functions, e.g. enthalpy, entropy, heat capacity, volume and compressibility, can be defined in a similar manner. By definition, excess thermodynamic functions are zero for $x = 0$ and $x = 1$. Also by definition, positive deviations from Raoult's law are equivalent to $G^{\mathrm{E}} > 0$, and negative deviations to $G^{\mathrm{E}} < 0$.

Liquid mixtures can be conveniently classified according to the signs and magnitudes of the excess functions. For mixtures of chemically similar and symmetrical molecules G^{E} is likely to be small, positive and symmetrical about $x = 0.5$. For molecules of dissimilar sizes, or where there are specific interactions such as hydrogen-bonding, the excess function diagrams can be quite complex, and nowhere is this more so than where one of the components is water.

Figure 6.1 illustrates several types of behaviour observed for aqueous and non-aqueous mixtures, although not all at the same temperature. Any comparisons of the different types of behaviour must be qualitative, because theories of concentrated mixtures of complex and dissimilar molecules do not yet exist. Nevertheless, several important features can be distinguished. The Gibbs free energy is a composite function with contributions from the energy (enthalpy) and the entropy. In several of the diagrams in Figure 6.1, H^{E} and TS^{E} are quite large and show marked

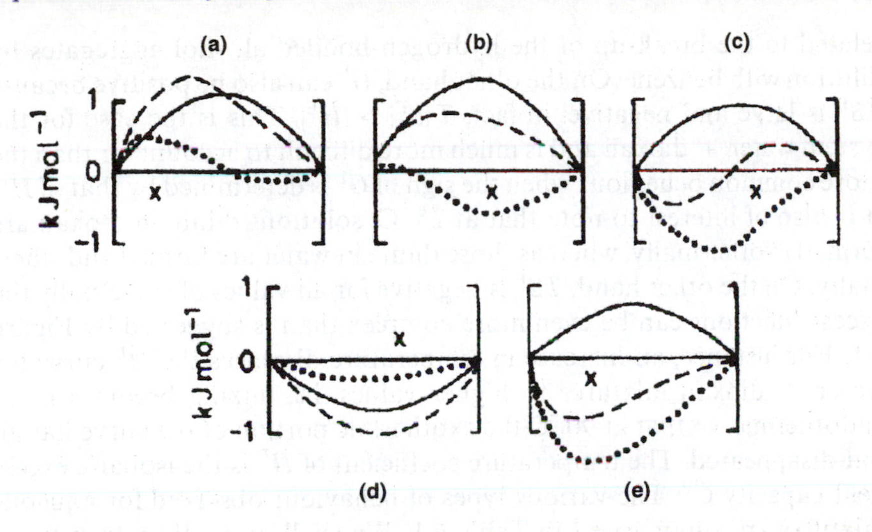

Figure 6.1 *Excess thermodynamic functions, as defined by equation (6.7), for the following mixtures (—— = G^E, ------ = H^E, ······ = TS^E):* (a) *ethanol/benzene;* (b) *methanol/benzene;* (c) *water/dioxan;* (d) *water/hydrogen peroxide;* (e) *water/ethanol; x is the mole fraction of the second component in each pair*

asymmetry about the *x*-axis. G^E, on the other hand, is remarkably symmetrical, irrespective of the nature of the two molecular species. This illustrates an important general rule: the free energy is not a very discriminating function, particularly for aqueous mixtures. G^E can usually be fitted to any but the most inapplicable model. Any eccentricities in H^E and TS^E seem to cancel out, giving rise to an apparently well-behaved, simple $G^E(x)$ relationship. This general rule also applies in biochemistry, where one commonly finds thermodynamic or spectroscopic free-energy data being used to support some hypothetical model mechanism, sometimes quite detailed and refined. More often than not, the experimental data are confined to equilibrium constants (which are also free energies, since $-\Delta G = RT\ln K$), determined at a single temperature. In view of the above discussion on the insensitivity of ΔG, such interpretations are likely to be of little value except as exercises in curve fitting. This often becomes apparent when attempts are made, unsuccessfully, to account for the temperature and/or pressure dependence of ΔG in terms of the postulated model. At times not even the sign of ΔH or ΔV can be correctly accounted for, despite the apparent good fit of ΔG.

Another feature of importance in Figure 6.1 is the numerical balance between the H^E and TS^E terms. Thus, G^E can be positive for one of two reasons: most frequently $|H^E| > T|S^E|$, in which case a positive G^E arises from endothermic mixing (as in the system ethanol + benzene). This is related to the break-up of the hydrogen-bonded alcohol aggregates by dilution with benzene. On the other hand, G^E can also be positive because TS^E is large and negative: in fact, $T|S^E| > |H^E|$. This is the case for the system water + dioxan and is much more difficult to account for than the more common behaviour when the sign of G^E is determined by that of H^E. It is also of interest to note that at 25 °C, solutions dilute in dioxan are formed exothermally, whereas those dilute in water are formed endothermally. On the other hand, TS^E is negative for all values of x. Actually the excess functions can be even more complex than is suggested by Figure 6.1. For instance, an increase in temperature displaces the H^E curve for water + dioxan mixtures to higher values, i.e. mixing becomes more endothermic, so that at 90 °C the exothermic portion of the curve has all but disappeared. The temperature coefficient of H^E is the isobaric excess heat capacity C^E. The various types of behaviour observed for aqueous mixtures are summarised in Table 6.1. We shall presently return to an examination of the origins of some of these effects.

While diagrams of the type shown in Figure 6.1 and classified in Table 6.1 demonstrate qualitatively the different types of behaviour encountered, quantitative information about intermolecular interactions between the components can be extracted only from the limiting slopes, as either x_1 or x_2 approaches zero. Thus, experimental measurements performed at very low concentrations make it possible to calculate partial molar excess properties at infinite dilution which, in turn, provide an estimate of solvent–solute interactions. A popular way of approaching solvation is by means of the so-called standard thermodynamic transfer

Table 6.1 *Classification of solutes according to their thermodynamic excess functions*

Solute	*Function*					*Phase behaviour*
	H^E	G^E	TS^E	V^E	C_p^E	
Nitriles, ketones	+	+	−	+	−	Complete miscibility or upper critical demixing
Ethers, alcohols, amines	+ or −	+	−	−	+	Upper critical demixing tending to closed loop
Sec-, *tert*-amines, cyclic ethers, polyethers	−	+	−	−	+	Closed loop
Urea, hydrogen peroxide, polyols, sugars	−	−	−	−	~0	Completely miscible

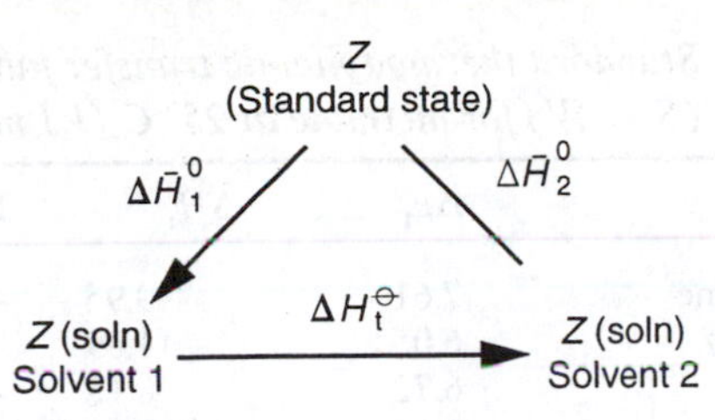

Figure 6.2 *Representation of the standard enthalpy of transfer of Z from solvent 1 to solvent 2; for explanation see text*

functions. The aim is to compare the behaviour of a given solute in a number of different solvent environments. This principle is illustrated in Figure 6.2, the enthalpy being used as the property under study. The heat of solution of solute Z in solvents 1 and 2 is measured as a function of concentration of Z, and the limiting partial molar enthalpy of solution is obtained, as $\Delta\bar{H}_1^0$ and $\Delta\bar{H}_2^0$. The difference, $\Delta H_t^{\ominus}$, known as the standard heat of transfer, corresponds to the heat liberated when one mole of Z is transferred from an infinitely dilute solution in solvent 1 to solvent 2. The reference state of Z can be either pure Z in the liquid state or the ideal gas. The two solvents can be either pure components or mixtures.

The principle of standard transfer functions can best be illustrated by some typical examples. Table 6.2 contains a summary of standard chemical potentials associated with the transfer of various probe solutes (at infinite dilution) from a variety of solvents (S) to water (W). In other words, $\Delta\mu_t^0$ is a measure of the difference in solvation interactions experienced by the probe molecule in the two solvent media. Thus, $\Delta\mu_t^0 < 0$ indicates that the probe molecule prefers the aqueous environment to solvent S, and vice versa for $\Delta\mu_t^0 > 0$.

The simple solutes included in Table 6.2 interact more favourably with

Table 6.2 *Free energy of transfer of solute from solvent to aqueous solution at 25 °C of simple solutes (kJ mol^{-1})*

Solute	*Solvent S*	$\Delta\mu_t\ (S \rightarrow W)$
Methane	Ethanol	6.72
Ethane	Ethanol	10.2
Methane	D_2O	0.13
Ethane	D_2O	0.15
Argon	n-Hexane	6.50
Argon	Ethanol	5.03
Argon	Ethanediol	0.21
Methane	1 M NaCl (aq)	−0.85

Table 6.3 *Standard thermodynamic transfer functions ($S \rightarrow W$) for methane at 25 °C (kJ mol^{-1})*

Solvent (*S*)	$\Delta\mu_t$	ΔH_t	$T\Delta S_t$
Cyclohexane	7.61	−9.95	−17.56
1,4-Dioxan	6.05	−11.89	−17.94
Ethanol	6.72	−8.18	−14.90
Methanol	6.68	−797	−14.65

solvent S than with water, with the single exception of methane in an aqueous NaCl solution. Since, according to Figure 6.2, $\Delta\mu_t$ is the difference of the free energies of solution, it is also a measure of the relative solubilities of the probe in S and in water. The results in Table 6.2 therefore demonstrate that methane, ethane and argon are more soluble in S than in water, except where S is an aqueous NaCl solution. The results further demonstrate that for a given solute, argon, $\Delta\mu_t$ becomes less positive as S becomes more polar for members of a homologous series with increasing molecular weight. It is of course easy to rationalise such results in terms of: 'oil and water don't mix'. Indeed, it is to be expected that a hydrocarbon would feel happier in hexane or ethanol than in water. Not quite so easy to account for is the observation that hydrocarbons are more soluble in D_2O than in H_2O.

Hydrophobic Hydration – Structure and Thermodynamics

On probing a little deeper into the reasons why $\Delta\mu_t^0$ is positive, we find that the above facile rationalisation is completely misleading. It is possible to measure the enthalpy contribution to $\Delta\mu_t$, either directly by calorimetry or indirectly by measurements of $\Delta\mu_t$ at several temperatures. Knowing the free energy and enthalpy, the corresponding entropy of transfer is then easily calculated. Table 6.3 summarises such data for the transfer of methane from several solvents to H_2O. Although $\Delta\mu_t$ is found to be positive, the interaction energy (enthalpy) is negative in all cases, i.e. exothermic transfer, demonstrating that the solvation energy for methane in water is *more* favourable than the corresponding energy for methane in the other solvents. The reason why methane is less soluble in water is therefore the large negative entropy change which accompanies the transfer of methane from any other solvent to water. The possibility that a positive free energy can arise from a large negative entropy term has already been mentioned (Table 6.1). This behaviour, hardly encountered with nonaqueous systems, is not easy to explain. It is, however, characteristic of aqueous solutions of apolar molecules and those more polar organic molecules that possess only *one* functional group: alcohols, monoamines, ketones and ethers. This observation led to intensive experimental studies of aqueous solutions of noble gases and hydrocarbons. For solutions of more polar compounds, e.g. H_2O_2, urea, carbohydrates, glycols and polyamines, the sign and magnitude of $\Delta\mu_t$ is dominated by the sign and magnitude of ΔH_t (see, for instance, the behaviour of $H_2O + H_2O_2$ mixtures, shown in Figure 6.1).

Since the ideal entropy of mixing must be positive for all values of x, such large negative excess entropies are not easy to account for. The most satisfactory way of rationalising the observed results is to postulate that the introduction of apolar molecules, or apolar residues on otherwise polar molecules, into water leads to a reduction in the degrees of freedom – spatial, orientational, dynamic – of the neighbouring water molecules. In other words, water becomes configurationally, more constrained. Originally it was suggested that apolar residues induce a quasi-crystallisation of water, and the term *iceberg* was used, but this has led to much confusion and misunderstanding. The effect is now called *hydrophobic hydration.*

While the model of an iceberg around a nonpolar molecule in solution may not be a realistic concept, there does, however, exist a class of crystalline structures that closely resembles this picture of hydrophobic hydration. Thus if solutions of noble gases, hydrocarbons and slightly polar molecules are cooled, then the solid phase that separates out does not consist of ice but of a gas hydrate in which water provides a hydrogen-bonded framework containing holes or cavities that are occupied by the solute (or guest) molecules (see Chapter 4). The framework resembles ice in the sense that every water molecule is hydrogen-bonded to four other water molecules, but the tetrahedral hydrogen-bond geometry characteristic of ice is somewhat distorted. Depending on the degree of tetrahedral angle distortion, cavities of different diameters can be created, capable of accommodating guest molecules of different dimensions. The smallest cavity is a pentagonal dodecahedron, shown in Figure 4.6. A common clathrate form combines these dodecahedra with tetrakaidecahedra, made up of twelve pentagonal and two hexagonal faces. These latter cavities are found in the hydrates of simple alkanes, ethylene oxide, tetrahydrofuran, carbon dioxide and sulfur dioxide. The guest molecules are generally not bonded to the water lattice but are free to rotate inside the cavities.

Thermodynamic, NMR and computer simulation studies of aqueous solutions of clathrate-formers suggest that the clathrate hydrate is a realistic model for the phenomenon of hydrophobic hydration. Thus, the observed negative TS^E values arise from the reduction in the allowed configurational degrees of freedom of water molecules in the vicinity of the apolar molecule and a reduction in the diffusion rates of such water molecules. Whilst for many years the suggestion of a clathrate hydrate structure in solution was based on circumstantial evidence, recent neutron diffraction measurements on argon solutions in water have confirmed the correctness of the earlier speculations. Thus the $g_{w\text{-}Ar}(r)$ places the Ar atom at the centre of a cavity composed of 16 ± 2 water molecules, with the first peak (Ar–O distance) centred on 0.35 nm. The geometry and coordination number thus closely resemble those of the clathrate hydrate cage. As would be expected, the 'solution structure' of the hydration shell is somewhat disordered, but the neutron data also show evidence of a second shell that extends to 0.8 nm, suggesting that hydrophobic hydration may be a long-range interaction. If true, this result would have an important bearing on calculations and simulations of solution properties, not least those associated with protein stability (see Chapter 10). It is not certain whether the range of such hydration is confined to the nearest-neighbour water molecules or how far it might extend. In other words, the shapes of $W_{12}(r)$ and $g_{12}(r)$ are as yet unknown; here 1 refers to the solvent (more correctly, the oxygen atom) and 2 to the solute which is assumed to be monatomic.

Other solution properties characteristic of hydrophobic hydration include negative limiting excess partial molar volumes $\bar{V}_2^E$, defined by $\bar{V}_2^E = \bar{V}_2 - V_2^0$, where V_2^0 is the molar volume of the solute in its pure liquid state, and positive excess partial molar heat capacities, $\bar{C}_2^E$, defined similarly. Both these observations are readily accounted for on the basis of the clathrate hydrate model. A negative $\bar{V}_2^E$ implies a net volume shrinkage when the solute is added to water. In other words, the solute in aqueous solution does not occupy as much volume as it does in its pure liquid state. This is due to the considerable empty space that exists in water anyway, because of its four-coordinated structure. All that is required on the introduction of a solute molecule is a slight reorganisation of this free volume to form the cavities capable of accommodating the solute molecule. It is relevant here to note that for a homologous series, $\bar{V}_2^E$ appears to be a linear function of the number of $-CH_2$ groups in the solute molecule. The same is true for the solute dependence of $\bar{C}_2^E$. The positive sign of $\bar{C}_2^E$ signifies that the introduction of the solute into water promotes 'structures' that are thermally more labile than is pure water; here again the clathrate model seems to fit the data.

Hydrophobic Interactions

Having dealt with the nature of the solute–water interaction, let us extend the discussion and enquire what happens when two (or more) such hydrated apolar molecules are allowed to interact. This process is known as the *hydrophobic interaction.* As was explained earlier, information about the interaction between pairs of molecules in solution is derived from the limiting concentration dependence of the various physical properties. For the excess free energies shown in Figure 6.1, we require the value of $\partial G^E/\partial x_2$, as $x_2 \to 0$. Formally, excess properties like the osmotic pressure, can be expressed in terms of power series in concentration. Thus, for the free energy

$$G^E = g_{22}m_2^2 + g_{222}m_2^3 + \ldots \tag{6.8}$$

where the virial coefficients $g_{22}, g_{222}, \ldots$ express the contributions to G^E by pair, triplet, etc. molecular interactions. It should be noted that g_{22} is here a virial coefficient in a free energy expression. (It must not be confused with $g_{22}(r)$, the solute–solute radial distribution function.) Similar equations can be written for all the other excess properties. If measurements are performed on dilute solutions, then only the first term in equation (6.8) contributes significantly to G^E.

Having seen that, in terms of the free energy, the introduction of a nonpolar molecule into water is a highly unfavourable process, we would expect that g_{22} in equation (6.8) should be negative. In order to minimise unfavourable water–solute interactions, solute molecules will interact preferentially, thus reducing the number of their solvent contacts. This is indeed found to be the case, and the experimental observation has given rise to the seemingly obvious conclusion that the hydrophobic interaction is a partial reversal of the entropically unfavourable hydration process. According to this picture the second virial coefficients, h_{22}, v_{22} etc. should all have the opposite signs to those of the corresponding properties associated with the hydrophobic hydration, such as H_2^{0E}, V_2^{0E} etc. For the free energy, enthalpy and entropy this is found to be the case. There is still some doubt as to the limiting concentration dependence of C_2^E, but the excess volume does not conform to the above model and this raises doubts about its validity. Figure 6.3 combines the partial excess volume data for a variety of solutes in aqueous solution. Although none of the solutes is 'simple' in terms of the definition given at the beginning of this chapter, most of them exhibit a dilute solution behaviour that corresponds closely to that of the 'simple' molecules (but see also Chapter 8). The limiting excess volumes are negative in each case, reflecting the open structure of water which is capable of accommodating solutes.

However, the concentration dependence of V_2^E allows the differentiation between so-called hydrophobic solutes and solutes that are predominantly polar. For the former class of compounds $\partial V_2^E/\partial x_2 < 0$,

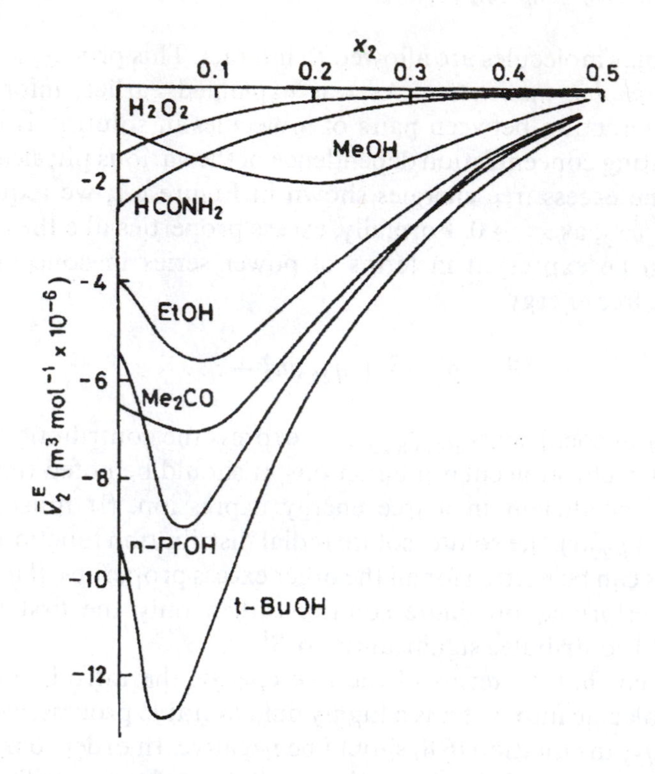

Figure 6.3 *Excess partial molar volumes, at 25 °C, of representative solutes in aqueous solution as function of solute concentration*
(Reproduced by permission from *Water – A Comprehensive Treatise*, ed. F. Franks, Vol. 4, Plenum Press, New York, 1975, p. 43)

whereas for the polar compounds this derivative is positive. This is also the type of behaviour one would expect if the hydrophobic interaction was simply a partial reversal of hydrophobic hydration, i.e. solute–solute contacts replacing solute–water contacts. The $\bar{V}_2^E(x_2)$ curves for the alcohol series are quite typical of the volumetric behaviour of hydrophobic solutes, demonstrating that, despite the presence of the –OH group, alcohols, even methanol, behave as 'soluble hydrocarbons' rather than 'alkylated water'.

The existence of minima in the curves indicates that the interaction responsible for the negative limiting $\bar{V}_2^{0E}$ values becomes more, not less, pronounced as the alcohol concentration increases, i.e. the shrinkage of the mixture per mole of added alcohol becomes more pronounced with increasing concentration, up to a point. In terms of the clathrate hydrate model this means that two adjoining water cavities are more stable than two isolated cavities when both are occupied by guest molecules. The effect is more pronounced for the larger solute molecules. Eventually the sign of $\partial V_2^E/\partial x_2$ reverses and $V_2^E(x_2)$ then approaches zero. Presumably at some given solute concentration there will no longer be enough water molecules left to provide a fully intact cavity structure. The alcohol molecules will then be able to interact directly, water being squeezed out and relaxing to its normal configuration state. At these concentrations, however, the higher order terms in equation (6.8) are quite important, and the minima in the $V_2^E\partial x_2)$ curves are therefore hard to interpret in any but qualitative terms. The data in Figure 6.3 refer to 25 °C; if the temperature and pressure dependence of V_2^E are examined, they are found to be extremely complex. This is particularly well illustrated by the effect of hydrophobic solutes on the temperature of maximum density (TMD) of water. It can be shown that *all* solutes should produce a TMD shift to lower temperatures, so that in terms of the excess functions previously defined in equation (6.7), the ideal solution contribution is negative. However, hydrophobic solutes produce a TMD shift which is either less negative than the calculated ideal value or which is actually positive. Table 6.4, which compares the thermodynamic properties of hydrophobic solutes with those of normal solutes in aqueous solution, highlights the fundamental differences.

Before leaving the thermodynamics of hydrophobic interactions it is necessary to point out that net attractive interactions between solute molecules are also observed in solvents other than water, and also for polar solutes in water. One can therefore define a general *solvophobic interaction*, the criterion of which is that $g_{22} < 0$. The essential differences between this general effect and the hydrophobic interaction, as previously discussed, are strikingly illustrated in Table 6.5 which compares the various pair interaction parameters, as defined in equation 6.8 for urea and ethanol. Urea is a typical polar solute, which forms well-behaved (as well behaved as aqueous solutions can ever be) mixtures with water. The negative g_{22} parameters for the two solutes are of the same order of

Table 6.4 *Limiting thermodynamic excess functions of hydrophobic and 'normal' solutes in aqueous solution* (ΔX_t^{0E})

Function	*Hydrophobic solutes*	*Polar (normal) solutes*
Free energy	+	−
Enthalpy	+ or −	−
Entropy	−, T|ΔS| > |ΔH|	−, T|ΔS| < |ΔH|
Heat capacity	+	−(∼0)
Volume	−	(small)
Expansibility (at low temp.)	−	+
Shift in TMD	+	−
Compressibility	− (small)	− (large for ions)

Table 6.5 *Thermodynamic pair interaction coefficients for ethanol and urea in aqueous solution at 25 °C [J mol^{-1} (mol kg^{-1})$^{-1}$]*

Coefficient	*Ethanol*	*Urea*
g_{22}	−94	−82
h_{22}	243	−351
Ts_{22}	337	−269

magnitude, indicating that in solution there appears to be a net attraction between pairs of solute molecules. As pointed out previously, the free energy often disguises more than it reveals. In this case it hides the fact that the net attraction arises from very different origins, with urea behaving in the normal, expected way, while ethanol exhibits the symptoms of the hydrophobic interaction.

The thermodynamic excess functions, both at infinite dilution and at finite concentration, suggest that the twin phenomena of hydrophobic hydration and hydrophobic interactions reflect structural changes in water, induced by the apolar molecules or residues, i.e. by chemical groups that are unable to participate in hydrogen-bonding. Modern statistical mechanical studies and computer simulation experiments have provided the first details as to the possible nature of such effects. One of the most important conclusions is that $W_{22}(r)$ is an oscillating function with several minima, unlike the simple Lennard–Jones 6–12 potential function which is frequently employed to describe intermolecular forces where only van der Waals interactions need be considered. Since the experimental evidence is overwhelming that hydrophobic effects are dominated by the entropic, configurational terms, it follows that some of the minima in $W_{22}(r)$ should become *deeper* with rising temperature, the solute–solute attraction becoming stronger. In the extreme case, the organic component may even separate out as a pure phase as the temperature is raised, a phenomenon known as lower critical demixing (see, for instance, the phase diagram of water–tetrahydrofuran, which exhibits a lower critical solution temperature at 72 °C).

Dynamic Manifestations of Hydrophobic Effects

The peculiar symptoms of hydrophobic effects can also be observed in the dynamic behaviour of aqueous solutions. A particularly effective method for studying the diffusional motions of solute and solvent in a mixture makes use of nuclear magnetic relaxation. The time decay of a previously induced nuclear magnetisation is measured, and this can be related to the rotational correlation time τ_c, as described in Chapter 3. The particular advantage of the NMR method is that, by appropriate isotopic substitutions, the relaxation behaviour of different nuclei, belonging to solute or solvent molecules, can be measured independently in the mixture. Figure 6.4 provides an example of this approach for the system water–acetone: the two upper curves describe τ_c of ^1H water only, because the solute was present in its fully deuterated form. The lower three curves show the rotational diffusion behaviour of the acetone molecule, measured both by ^1H and ^{17}O magnetic relaxation. In this case the solvent was D_2O. To obtain even more detail of the diffusional motions of the acetone molecule, additional measurements could be performed on ^{13}C.

Five important conclusions can be drawn: (i) the initial addition of acetone to water leads to a slowing down of the rotational diffusion of water molecules, but, as the acetone concentration is increased, the effect gradually disappears. (ii) This inhibition of the rotational motions of water is very sensitive to temperature. (iii) Hardly any corresponding slowing down of the acetone molecules is observed; in fact, acetone, despite its larger size, rotates more rapidly than do the water molecules. (iv) Only a very slight temperature effect is observed for the rotational correlation time of acetone. (v) The similarity of the values of ^1H and ^{17}O

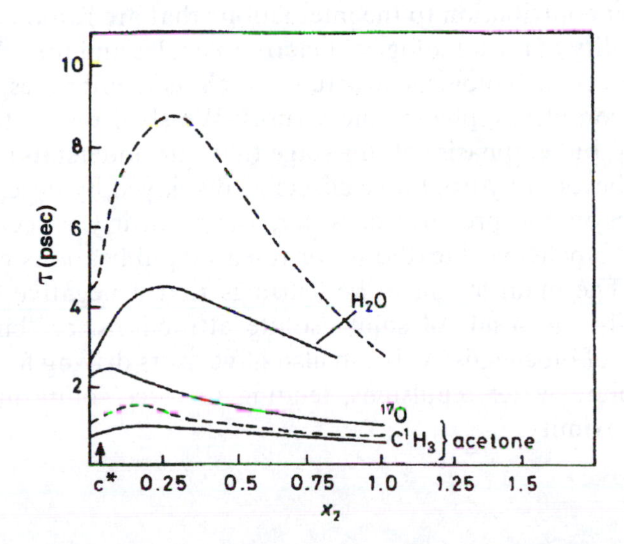

Figure 6.4 *Rotational correlation time τ of ^1H and ^{17}O in water and acetone (as indicated) as function of acetone concentration. Broken lines refer to 5 °C and solid lines to 25 °C. The concentration c* corresponds to the acetone clathrate hydrate stoichiometry acetone. 17H_2O*
(Reproduced by permission from *Water – A Comprehensive Treatise*, ed. F. Franks, Vol. 4, Plenum Press, New York, 1975, p. 23)

magnetic relaxation rates in acetone suggests that the acetone molecule in aqueous solution performs isotropic tumbling motions.

The following picture emerges: when acetone is added to water, the diffusional motions of water molecules are inhibited, but there is no corresponding inhibition to the motions of acetone. The activation energy for the rotational diffusion of water is large, but that of acetone is close to zero. The values for the acetone diffusion lie close to the values associated with the diffusion of acetone in its dilute vapour state or as a guest molecule in a crystalline clathrate hydrate. All these findings are completely compatible with the cavity model of the hydrophobic effect; compare also the shape of the τ_c curve for water with the $V_2^E(x_2)$ curves in Figure 6.3. The composition denoted by c^* and the arrow in Figure 6.4 corresponds to the stoichiometric composition of the acetone clathrate hydrate. This suggests that the actual structures in aqueous solutions of hydrophobic solute species are fairly rudimentary and incompletely formed, some of them requiring a smaller number of water molecules than does the stoichiometric crystalline clathrate. This conclusion is again consistent with computer simulation studies of the hydrophobic interaction.

The above discussion of hydrophobic effects has been so detailed because the hydrophobic interaction is now recognised as providing a dominant contribution to the interactions that are responsible for the *in vivo* stability of most biological macromolecules and for self-assembly of supermolecular biological structures, such as membranes, multisubunit enzyme complexes, phages and viruses. We shall return to this subject presently, but emphasise at this stage that our understanding of some of the attributes of hydrophobic effects as displayed by deceptively simple molecules in the presence of water, is not nearly as certain as most standard biochemical textbooks or research publications would make it appear. The main lesson to be learnt is that a negative $W_{22}(r)$ is not necessarily the result of solute–solute attractions, or 'binding' in the language of biochemistry. It can also have as its driving force the sum of many solute–water repulsions, tending to force solute molecules into closer proximity.

Chapter 7

Aqueous Solutions of Electrolytes

Classical Electrostatics

In principle, a molecular description of an ionic solution is no different from that of a solution of a nonelectrolyte. In practice, the problems are of a different type. In the first place an ionic solution will, of necessity, contain at least two solute species: a cation and an anion. The solute–solute interaction potential $\mu_{+-}(r)$ and radial distribution function $g_{+-}(r)$ therefore contain at least three contributions. On the other hand, there are many monatomic ions, so that interactions can be treated as spherically symmetric, a poor approximation for many nonelectrolytes (see the discussion on polar molecules in the following chapter). A simplifying factor in the case of electrolytes is that a dilute solution law exists that enables extrapolations to infinite dilution to be performed with some degree of confidence. In other words, long-range effects are well accounted for by electrostatics, where the interactions are proportional to r^3. The Debye–Hückel limiting law provides information about the ion distribution around a given ion. This is obtained by first solving the Poisson equation for the cation–anion potential as a function of the distance of separation, $u_{+-}(r)$, which is then inserted in the Boltzmann equation to calculate the distribution function. Furthermore, it is assumed that the excess free energy ΔG^{E}, as defined by equation 6.7 arises solely from electrostatic interactions and can therefore be calculated with the aid of the Debye–Hückel law. This is found to be quite satisfactory for large values of r (i.e. low concentrations). The existence of a solvent is acknowledged only via its macroscopic dielectric permittivity.

At higher concentrations the weaknesses of a purely electrostatic treatment begin to cause problems. Two methods exist for coping with such situations: (i) the simple solution laws continue to be used, modified by the addition of empirical or semi-empirical parameters to account for deviations from predicted behaviour; (ii) a new molecular model is developed that allows the *de novo* calculation of solution properties at higher ion concentrations or in mixed solvents. In practice, the first approach has been preferred by most workers in the field, i.e. the extension of the Debye–Hückel equation, by adding 'correction' terms that include extra-electrostatic constants, such as ionic radii and ion association constants. These devices are successful for ionic solutions in nonaqueous solvents, but the peculiarities of water and some aqueous solutions produce some quite remarkable anomalies, e.g. that the sum of the calculated cation and anion radii in solution is less than the sum of their atomic radii.

There grew a general feeling among electrochemists of the early 1970s that the time was ripe for a reassessment of established treatments and for the development of a new generation of more refined electrolyte theories which would take account of the molecular structure of the aqueous solvent environment.

Short-range Effects – Ion Hydration

Clearly, the above approach is unsatisfactory where short-range forces become of importance. Nor is it able to provide information about the close environment of an ion (solvation). That one or several molecular layers of solvent are markedly affected by an ion is now beyond doubt, and the question is often asked what might the dielectric permittivity of the solvent be when its molecules have been reorganised in the ionic solvation sphere, as a result of ion–dipole interactions. This question illustrates the dilemma of the solution chemist and the biochemist; reduced to its ultimate, it is: what is the dielectric permittivity of a water molecule? It is impossible to reconcile a molecular description with distribution functions and bulk properties that have been arrived at by various averaging processes involving large numbers of molecules.

There has for some time been a general agreement that the description of ionic solutions in terms of simple electrostatic potential functions modified by a bulk dielectric permittivity is no longer adequate. This had led to the birth of the second generation of theories of electrolyte solutions. Simultaneously, new experimental methods have made possible the direct probing of the ionic hydration shell, and the results obtained from such studies can be fed back into the newly developed theories in efforts to improve them. In the following paragraphs some of those developments will be briefly reviewed.

The combination of equations (6.5) and (6.6) yields the osmotic second virial coefficient of a solution in terms of the various solute–solute pair potential functions $u_{ij}(r)$, where i and j are the two solute particles, i.e.

cation and anion. In the simplest case where the ions are regarded as hard spheres, the potential has the form shown in Figure 7.1(a). Note that the electrostatic part of $u_{ij}(r)$ has been omitted here; we are interested solely in any contribution due to solvation

Now let us assume that each ion can strongly affect one molecular layer of solvent of thickness d^*, but that otherwise the potential function is unchanged. The resulting $u_{ij}(r)$ may then have the appearance shown in Figures 7.1b or 7.1c. A_{ij} is the free energy required to disrupt the solvation layers as the solute particles are brought together. Since by definition the solvation sphere is limited to one molecular layer, the repulsive contribution of solvation to $u_{ij}(r)$ is of a short-range nature. We must also consider the possibility that solvation produces a net attraction, and that is illustrated in Figures 7.1(d) and 7.1(e). The hydration changes that take place during the approach of two hydrated ions are shown pictorially in Figure 7.2. The simple model of the monomolecular solvation layer may be elaborated by including two layers of solvent molecules, as shown in the lower model. Such subtle modifications may be detected by spectroscopic methods, but they do not affect the principles illustrated in Figure 7.1.

In order to quantify A_{ij}, it is necessary to construct a model ion–ion potential. One such potential function, which enjoys some popularity, is

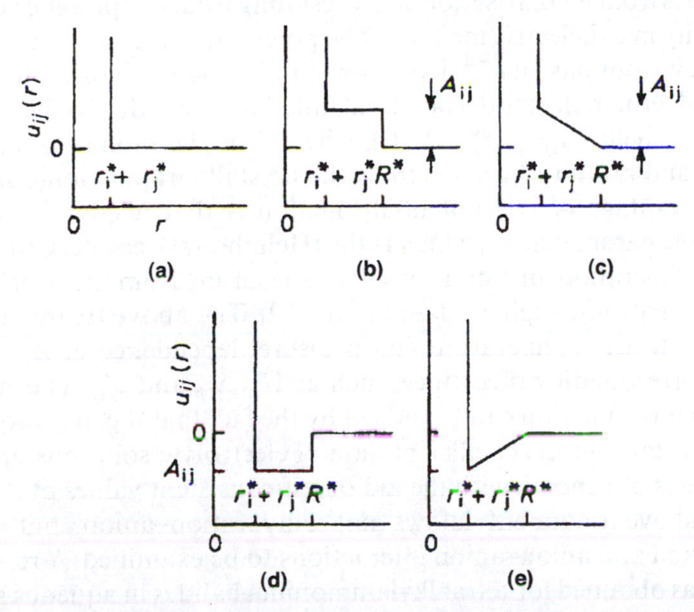

Figure 7.1 *Solute–solute pair potentials (excluding electrostatic contributions) for different solvation models;* $R^* = r_i^* + r_j^* + 2d^*$, *where* d^* *is the diameter of a solvent molecule. For explanation, see text*
(Reproduced by permission from *Water – A Comprehensive Treatise*, ed. F. Franks, Vol. 3, Plenum Press, New York, 1973, p. 16)

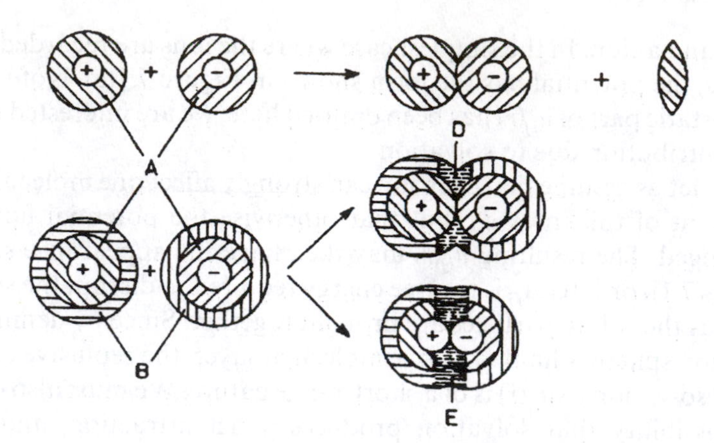

Figure 7.2 *Ion–solvent interaction in solution: A, primary hydration shells; B, secondary hydration shells; D, outer cosphere overlap (solvent-separated ion pair); E, primary cosphere overlap (contact ion pair)*

based on the sum of four contributions: an electrostatic term, given by $e^2/\varepsilon r$, where e is the electronic charge; a short-range repulsion with an exponential or power law dependence on r; and a weakly attractive term that arises from a polarisation effect resulting from the presence of the ion in a cavity in a dielectric medium. The potential of the ion–cavity interaction of two ions has an r^{-4} dependence and is therefore of medium range. The final term in the model potential function is that due to the overlap of solvation shells: $A_{ij}(r_i^*, r_j^*, d^*, V)$, where V is the molar volume of the solvent and r_i^* and r_j^* are taken as the crystallographic ionic radii. The great advantage of this potential function is that it contains only *one* adjustable parameter, A_{ij} which is the (Helmholtz) free energy associated with the 'liberation' of 1 mole of solvent, resulting from the overlap of the two ionic solvation spheres (see Figure 7.2). The above treatment can be extended to the temperature and pressure dependence of A_{ij}, and this yields corresponding quantities, such as U_{ij}, S_{ij}, and V_{ij}. The success of this potential function can be judged by the fact that it can reproduce the experimental osmotic coefficient data of electrolyte solutions up to concentrations of 1 molal, with the aid of self-consistent values of A_{ij}.

The above treatment allows not only cation–anion, but also cation–cation and anion–anion interactions to be examined. A remarkable result was obtained for tetraalkylammonium halides in aqueous solution: A_{++} was found to be negative, indicating that the solvation contribution to $u_{++}(r)$ was *attractive*. Furthermore, the corresponding entropy term, TS_{++}, was found to be positive, and V_{++} negative. These results are reminiscent of the hydrophobic interaction. It was also found that the

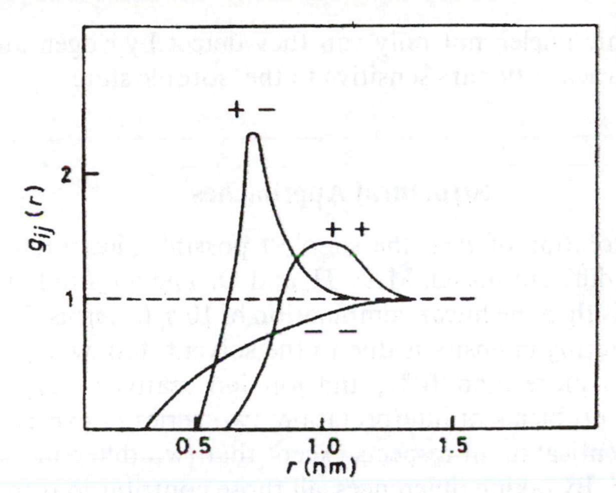

Figure 7.3 *Solute–solute (ion–ion) radial distribution functions for $Et_4N^+Br^-$ in aqueous solution, calculated from measured activity coefficients and the solvation shell overlap model*

magnitudes of the various solvation parameters increase with the number of carbon atoms in the alkyl residue. The three radial distribution functions for Et_4NBr are shown in Figure 7.3. Apart from the sharp peak of $g_{+-}(r)$, which is to be expected from electrostatic considerations, there is a well-developed peak in $g_{++}(r)$, indicating cation pairing. Since such a result could never be predicted by electrostatics, the Debye–Hückel theory cannot account for the aqueous solution behaviour of tetra-alkylammonium salts, where hydrophobic effects appear to provide an extra attractive contribution to the interionic forces.

The above arguments apply equally well to other types of ions which combine a charge with a substantial alkyl or aryl residue, such as long-chain alkyl sulfates (surfactants, acidic phospholipids, dyestuffs), etc. In all these cases massive and stable aggregation takes place of ions that carry *like* charges. Although a part of the total charge is neutralised by bound counterions, the net driving force to aggregation is the hydrophobic interaction (i.e. a large positive entropy change). The tetraalkylammonium halides have been, and are still being, extensively used as probes for studies of the hydrophobic interaction, because they are very water soluble, symmetrical and monovalent.

Just as the past two decades have witnessed a re-examination of classical electrolyte theories, experimental scattering techniques have been perfected that have made possible quite detailed studies of ionic solvation shells. The advent of neutron scattering methods has led to spectacular developments. The chief advantage is that neutrons interact with the atomic nuclei; not only can they detect hydrogen atoms, but scattering characteristics are sensitive to the isotopic state.

Structural Approaches

An aqueous solution of even the simplest possible electrolyte (M^+X^-) contains four different nuclei: M, X, H, and O. The *total* radial distribution function is then the linear combination of 10 $g_{ij}(r)$ terms. Most of the observed scattering intensity is due to the solvent. Ion–water scattering contributes no more than 10%, and ion–ion scattering even less. To overcome the problems of interpretation, two series of experiments are performed, identical in all respects except that two different cation isotopes are used. By taking differences, all those contributions to the total $g(r)$ that do not contain the cation cancel out, and we are left with only four out of the original ten terms: M–O, M–H, M–M and M–X. (Note: for various practical reasons, D_2O, rather than H_2O, is used in these experiments, but that does not affect the structural interpretation of the results.)

The results for a 1.5 molal solution of $NiCl_2$ in D_2O are shown in Figure 7.4. The first peak corresponds to the Ni–O distance and the

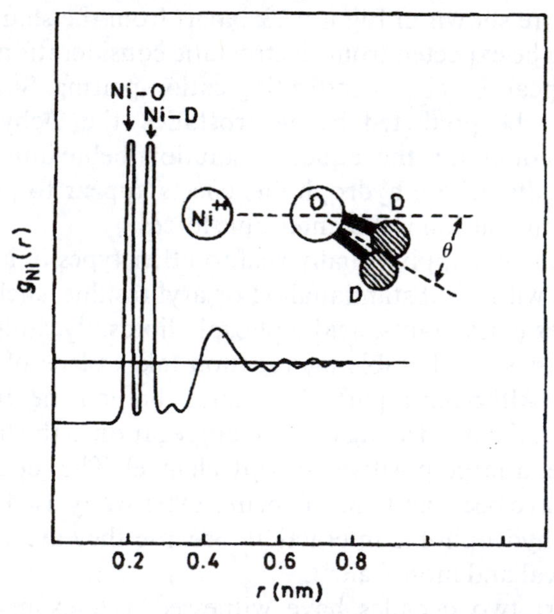

Figure 7.4 *Ni^{2+} radial distribution function, calculated from neutron scattering data on a 1.46 M solution of $NiCl_2$ in D_2O. The model shows the hydration geometry of Ni^{2+}*
(Reproduced by permission from *Water – A Comprehensive Treatise*, ed. F. Franks, Vol. 6, Plenum Press, New York, 1979, p. 29)

second peak to the Ni–D distance; the ratio of the areas under these peaks is of course 1:2, and integration of the peaks yields a hydration number of 5.8 ± 0.2 which is independent of concentration. This fact, taken together with the sharpness of the peaks, reflects the stability of the first hydration shell of the Ni^{2+} ion. From the positions of the two peaks and a knowledge of the geometry of the D_2O molecule it becomes apparent that the D_2O molecules are tilted at an angle to the Ni–O axis; this angle increases with increasing concentration, so that the hydration sphere is sensitive to distortion as the packing fraction of hydrated ions increases. In very dilute solutions this angle tends to zero.

The broad peak, centred on 0.45 nm, could be due either to Ni–Cl or to the second hydration shell, i.e. Ni–O. Figure 7.4 clearly illustrates the potency of the neutron scattering method.

Similar experiments have been performed in which isotope substitution has been applied to the anion, in this case Cl^-. The corresponding picture that emerges for the chloride ion hydration shell is that, once again, the hydration number is approximately 6, and that the D_2O molecules are disposed around the anion as shown in Figure 7.5. In no case can any structure be detected beyond 0.5 nm.

Experiments on other electrolyte solutions have so far revealed that the Ca^{2+} and Na^+ ions are surrounded by hydration shells closely resembling that of Ni^{2+}, the main difference being that the peaks are not as sharp. A surprising result, and one that conflicts with other circumstantial data, is that the Cl^- hydration shell is independent of the nature of the cation: Ca^{2+}, Na^+ or Rb^+. Not much can be said yet about cation–cation distributions, but it is claimed that the experimental results at reasonable concentrations clearly show that the Cl^- ion does not penetrate the cation primary-hydration shell.

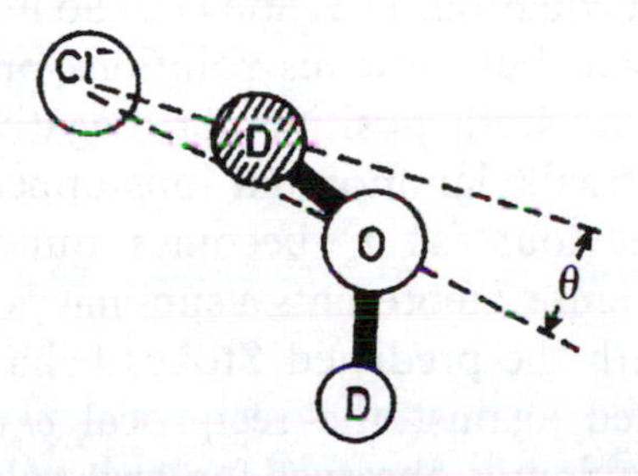

Figure 7.5 *Hydration geometry of the Cl^- ion hydration shell, obtained as shown in Figure 7.4*
(Reproduced by permission from *Water – A Comprehensive Treatise*, ed. F. Franks, Vol. 6, Plenum Press, New York, 1979, p. 31)

Assignment of Ionic Radii

There are still major problems with quite simple concepts, e.g. the correct assignment of ionic radii. This is of importance not only in studies of the transport properties of electrolytes in solution but in biochemistry and physiology where calculations involving the diffusion of ions, active or passive, and permeation of ions through barriers such as membranes always require knowledge of the radius of the diffusing species. Without wishing to labour the point, let us review briefly the difficulties relating to the correct assignment of ionic radii.

For the migration of an ion under the influence of an electric field we have the simple Stokes law:

$$\frac{\lambda_0 \eta}{Z} = \frac{Fe}{6\pi r_s} \tag{7.1}$$

where Z is the valence, λ_0 is the limiting ionic conductance, η is the viscosity, F is the Faraday and r_s is the hydrodynamic, or Stokes radius of the ion. Equation (7.1) describes the case where the solvent adheres perfectly to the surface of the (spherical) ion, i.e. perfect sticking. If the sphere is assumed to move without any adhesion to the solvent (perfect slipping) the equation takes the following form:

$$\frac{\lambda_0 \eta}{Z} = \frac{Fe}{4\pi r_s} \tag{7.2}$$

Inherent in the model is the condition that r_s must be large compared to the radius of the solvent molecules, but this is hardly ever the case in practice. Nevertheless, Stokes' law is used extensively to estimate ionic radii.

It is well known that equations (7.1) and (7.2) do not fit the experimental data for any solvent, but aqueous solutions present an especially complex behaviour, with both positive and negative deviations being observed. Only the tetraalkylammonium ions appear to approach the predicted Stokes behaviour, as r_s becomes quite large (e.g. tetra-*iso*amylammonium). Figure 7.6 presents a summary of some of the available data, together with the predicted Stokes behaviour. The Walden product, $\lambda_0 \eta$, is plotted against the reciprocal *crystallographic* ionic radius. Stokes law behaviour is observed for large values of r_x; eventually $\lambda_0 \eta$ passes through a maximum, and this has been rationalised in terms of a solvent dipole relaxation effect. It is postulated that a moving ion can orient the solvent molecules in the solvation sphere, and since such

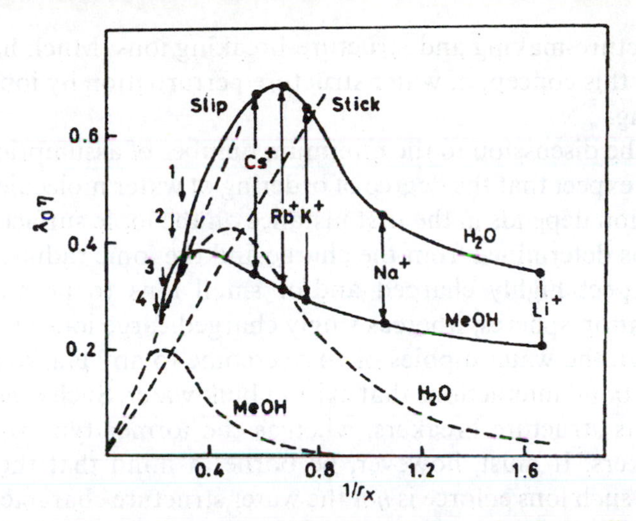

Figure 7.6 *Limiting Walden product* ($\lambda_0\pi$) *for alkali metal and tetraalkylammonium ions in water and methanol; numbers 1–4 represent tetramethyl, ethyl, propyl and butylammonium respectively. The broken lines represent the theoretical values of* ($\lambda_0\eta$)*, as explained in the text*
(Reproduced by permission from *Water – A Comprehensive Treatise*, ed. F. Franks, Vol. 3, Plenum Press, New York, 1973, p. 182)

oriented dipoles require a finite time to relax to their original unperturbed orientations after passage of the ions, a net retardation will be imposed on the ion. This effect has been quantified and the theoretical curve is included in Figure 7.6. Although the addition of a solvent relaxation effect now makes it possible to account for the shape of the Walden product dependence on r_x^{-1}, the *maximum* calculated value for λ_0 in aqueous solution is 29.4, whereas the observed conductances of several ions are as high as 75. Furthermore, the predicted maximum value of r_x is not in agreement with experiment, even when the correction for 'slippage' is applied. The best agreement between theory and experiment is observed for polar, aprotic solvents; the level of disagreement increases with the degree of association of the solvent.

The Concept of Structure Making/Breaking

In order to account, even only qualitatively, for the experimental results, the concept of overlapping hydration spheres must again be invoked. The hydration spheres are postulated to affect the viscosity of the solvent, and hence the ionic mobility. In the earlier discussion of inter-ionic forces it had been assumed that such hydration shells had a thickness of one water molecule, but such a simplified picture is not adequate to explain the ionic transport behaviour. We must now distinguish between so-called structure-making and structure-breaking ions. Much has been written about this concept of water structure perturbation by ions, some of it misleading.

Confining the discussion to the minimum number of assumptions, it is reasonable to expect that the degree of ordering of water molecules in the vicinity of an ion depends in the first instance on the ionic surface charge density. This is determined from the charge and the ionic radius, so that one might expect highly charged and/or small ions to possess well-ordered hydration spheres, whereas singly charged, large ions would not be able to align the water dipoles or to overcome to any marked extent the hydrogen-bond interactions that exist in bulk water. Such ions would be classified as structure breakers, whereas the former type would be structure makers. It must, however, be borne in mind that the actual structure that such ions enforce is *not* the water structure characteristic of the water radial distribution function in Figure 5.1, which reflects the tetrahedral hydrogen-bonding pattern in water. Rather, it is a structure imposed on the water by strong ion–dipole forces. The structure in question produces spherically symmetric orientations of water molecules, quite incompatible with the tetrahedrally hydrogen-bonded network in unperturbed water. Whereas one, two, or possibly even more water layers will perhaps be so oriented by a structure-making ion, there must also be a region where the magnitudes of the transmitted ion–dipole effects and hydrogen-bonding between water molecules become equal, and this must lead to a structural mismatch where the water molecules are structurally and dynamically highly perturbed.

From the preceding arguments one can deduce that structure-making ions might include Li^+, Mg^{2+}, F^- and polyvalent metal ions, whereas Rb^+, Cs^+ and I^- would be structure breakers. It is not so easy to classify polyatomic ions, but experience shows that SO_4^{2-} and PO_4^{3-} are powerful structure promoters, whereas NO_3^-, ClO_4^- and CNS^- are structure breakers. The hydration sphere hypothesis also implies that the ionic radius that determines the transport behaviour is not the crystallographic radius, but a quantity much less well defined.

The tetraalkylammonium ions $R_4N^+X^-$ appear once again to be a special case. Because of their hydrophobic nature they were believed to affect water differently, with the structure-making akin to the clathrate type water organisation. Indeed, many alkylammonium salts actually form crystalline clathrate hydrates (e.g. $(C_5H_{12})_4NF.38H_2O$). The hydration sphere, and hence the apparent ionic radii, are therefore affected differently. This then was believed to give rise to a more conventional r_x^{-1} dependence of $\lambda_0\eta$, as illustrated in Figure 7.6.

Once again, neutron scattering measurements have been able to test earlier speculations, based on circumstantial, non-structural evidence. Taking the tetramethylammonium ion (TMA) as the simplest possible test case, isotope substitution of N and of H in both TMA and water has revealed the necessary information on TMA–solvent, TMA–TMA and solvent–solvent correlations.

A particularly useful feature of the technique is the ability to check its internal consistency, provided that the molecular structure of the solute is known. Thus, the first peak in $g_N(r)$ must correctly describe not only the distance between the central nitrogen atom and the four methyl carbon atoms but also the number of such nearest-neighbour carbon atoms (4 in the case of TMA).

The next peak then refers, in a similar manner, to N–H distances, with a coordination number 12. Similar tests for self-consistency can be performed for the H_2O (D_2O) molecule.

The results suggest the following picture: the water environment of TMA is composed of approximately 20 water molecules, consistent with the expected number for a perturbed hydrophobic hydration shell, but there is no evidence for a special ordering of water molecules over and above that which is characteristic of pure water. This is not surprising, since the $g_{ij}(r)$ values for neighbour water molecules in ice and a clathrate crystal are closely similar. Differences are observed for pairs of molecules that are not hydrogen-bonded, i.e. next-nearest neighbours and beyond. It is instructive to note that the behaviour characteristic of hydrophobic residues disappears if the alkyl groups are modified so as to include $-OH$ groups. For instance, $(OHC_2H_4)_4N^+$ might be expected to possess a lower $(\lambda_0\eta)$ than $(C_3H_7)_4N^+$ because of its ability to form hydrogen bonds. Instead, $(\lambda_o\eta)$ is higher than that of the tetrapropylammonium ion, with which it is isoelectronic, by some 20%. Also, its effect on the viscosity of water is marginal, whereas the alkylated ion enhances η markedly. Observations of this type demonstrate the subtleties of hydrophobic effects when compared to the better-studied ion–dipole or hydrogen-bonding interactions. The temperature sensitivity of the ionic hydration sphere is expressed by the so-called viscosity B coefficient, defined by

$$\frac{(\eta - \eta_0)}{\eta_0} = Ac^{1/2} + Bc \qquad (7.3)$$

where η is the viscosity of a solution of concentration c, and η_0 is the viscosity of the solvent. In equation (7.3) the first term on the right-hand side accounts for electrostatic effects, and the constant A can be estimated from a Debye–Hückel type calculation. The second term arises from the short-range ion–ion interactions and from the overlap of ionic solvation spheres. At 25 °C the constant B is positive for structure-making ions and negative for structure-breaking ions. A plot of the temperature coefficient of B against that of the Walden products ($\lambda_o\eta$) has the form shown in Figure 7.7, whereas for the same ions in methanolic solution dB/dT is zero over the range 10–40 °C. Figure 7.7 demonstrates that water structural effects are very sensitive to temperature, i.e. the greater the ionic propensity to promote structure, the more negative dB/dT.

Water Dynamics

Structure making and breaking by ions has also been studied by various dynamic methods. In particular, the lifetimes of water molecules in the ionic hydration shell have been examined by NMR and dielectric

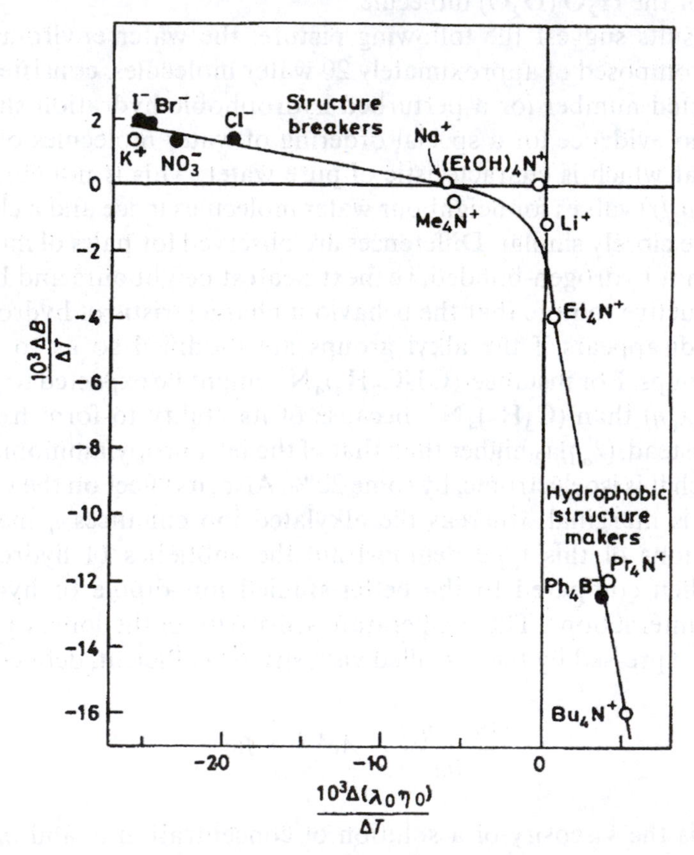

Figure 7.7 *Temperature dependence of the viscosity* B-*coefficient as function of the temperature dependence of the limiting Walden product* $\lambda_0\eta$
(Reproduced by permission from *Water – A Comprehensive Treatise*, ed. F. Franks, Vol. 3, Plenum Press, New York, 1973, p. 196)

methods. Thus in pure water at 25 °C the average lifetime of the water molecule at a given site is in the order of 1 ps. The lifetime of a water molecule in the hydration shell of a Mn^{2+} ion is 25 000 ps, whereas in the hydration shell of an I^- ion it is only marginally affected (4 ps). Intermediate values have been estimated for the hydration shells of other ions, for instance Na^+ (10 ps) and Cl^- (5 ps). The diffusion rates of water molecules *within* the hydration shells are also affected. In the vicinity of an I^- ion, water rotates more rapidly than it does in bulk water, whereas the opposite effect is observed for structure-making ions, where the rotational diffusion is retarded. From such dynamic data it is also possible to estimate the approximate height of the energy barrier to rotation of water, relative to the energy barrier in pure water: I^- 0.54, Cl^- 0.96, F^-, Na^+ and Li^+ 2.0.

Ion-specific Effects – Lyotropism

The foregoing discussion of ions in aqueous solution has established that long-range (low-concentration) properties can be adequately accounted for by electrostatics, taking into account only the magnitude of the ionic charge, the concentration and the dielectric permittivity of the solvent medium. This formalism breaks down at short range (high concentration), where specific ion–solvent effects dominate the solution behaviour. This distinction is of particular importance in biochemistry, where it is often found that the effects of electrolytes on certain processes are nonspecific up to concentrations of about 0.1 molal and can be expressed in terms of the ionic strength. At higher concentrations ion-specific effects become operative, although in the published literature this is not always made clear.

In this connection, an observation first made in 1888 is of overriding significance: in that year Hofmeister discovered that certain salts enhance the solubility of proteins in water, whereas others precipitate proteins. He was able to show that the effect was mainly due to anions and to establish a ranking (the Hofmeister or lyotropic series) of anions in the order in which they affect protein solubility:

(*precipitate*) SO_4^{2-}, HPO_4^{2-}, F^-, Cl^-, Br^-, NO_3^-, I^-, CNS^-, ClO_4^- (*solubilise*)

Although he was unable to explain the observations, he speculated that they were probably related to the different affinities of ions for water. Our understanding of the origin of the Hofmeister series has not advanced much since the original discovery, but it is now realised that this series appears in many different guises. Thus, the effect of electrolytes on the solubility of simple, nonpolar compounds in water follows the same series; sulfates salt-out, and perchlorates salt-in, argon or benzene in aqueous solution. Such effects cannot, of course, be accounted for on the basis of electrostatics. Here again, the tetraalkylammonium ions exhibit an atypical behaviour. They appear to enhance the solubility of all apolar compounds, the effect increasing with the size of the alkyl residue. Once again, the salting-in effect (which is a free energy function) originates from a favourable entropy change, rather than from a preferential energetic interaction between the ion and the other solute.

Ionic Solutions at Extreme Temperatures and Pressures

Recent years have witnessed the extension of thermodynamic and spectroscopic studies into the realm of high pressures and temperatures, beyond the critical point of water. Such data are of importance in several fields of technology, e.g. high-energy power plants and geothermal studies. Essentially the results have not shown up any major anomalies that might be associated with the passage from subcritical to supercritical conditions. Measurements of dilute solutions in the temperature range 603–674 K and pressures of 15–28 MPa (i.e. a water density of 200–650 kg m^{-3}) indicate no critical effects that would invalidate the Debye–Hückel Limiting Law. Interpolation equations, developed for ionic behaviour at moderate conditions, appear to apply equally well at 700 K and 30 MPa. The group additivity methods for estimating ionic solution properties at normal pressures and temperatures apply with the same degree of accuracy, although a few anomalies have been cited. For instance, apparent molar volumes of CsBr in dilute solution and at low densities (<350 kg m^{-3}) are low, i.e. less negative, compared to salts with smaller ionic radii. Walden's rule is not obeyed at the highest densities, where the calculated ionic radii apparently increase by 70%, possibly an indication of an increase in the number of water molecules 'carried' by ions.

The Hydrated Proton

In any discussion of processes that occur in aqueous solution, the nature and the participation of the hydrated proton be central. It is therefore hardly surprising that the proton in association with water has been subjected to many different experimental studies by a wide variety of techniques. The most convincing results have, however, been obtained from studies of the gaseous and solid states, while it is of course the liquid state where most practical interest resides. Recent years have witnessed a major shift to theoretical studies of various aspects of the hydrated proton, mainly the species H_3O^+; they include numerous molecular orbital (MO) calculations and computer simulations. Results have then been used to test the interpretations of previous experimental studies.

Theoretical studies of higher hydrates are fewer in number. For computational reasons, limitations have had to be imposed in most of the calculations. For instance, a species $H_3O^+(H_2O)_n$, with $n = 0$–4, was treated with certain symmetry constraints which approximated to H_3O^+ with two symmetrically attached water molecules of O··O lengths 246 pm. It was found that this configuration was significantly more stable (by 8.7 kJ mol^{-1}) than a structure consisting of $H_5O_2^+$ with one attached water molecule. Theoretical approaches have also been used to account for the remarkable stability of so-called 'magic numbers' in hydrated proton clusters of the type $H^+(H_2O)_{21}$ and $H^+(H_2O)_{28}$, detected in the gas phase.

Monte Carlo simulations are better than MO calculations at taking into account the solution environment of the hydrated proton. Thus, atom–atom pair potentials for a H_3O^+/H_2O interaction were first calculated by *ab initio* methods and then used in a simulation that included the hydrated proton and 215 water molecules in a cube of side 1860 pm. After equilibration, the $H_3O^+ \cdot\cdot H_2O$ interaction energy, averaged over the first hydration shell, was found to be -105 ± 0.8 kJ mol^{-1}. The geometry of the primary hydration shell suggests that four water molecules are involved, of which one molecule has to compete for hydrogen-bonding to one of the three hydrogen atoms of the ions. This is a novel feature which is, however, consistent with the experimental (X-ray and neutron diffraction) radial distribution functions. Representative results of this geometry are shown in Figure 7.8. The Monte Carlo radial distribution functions also provide evidence for at least two distinct hydration shells.

In any experimental study of the hydrated proton in an aqueous solution, account must be taken of the interaction of the ionic species with the surrounding medium. Credible interpretations of spectral and other data should also take account of proton exchange with counterions and/or solvent molecules, although such aspects are frequently circumvented in published reports.

Most structural studies have been confined to X-ray diffraction in acid solutions which is dominated by oxygen scattering. Neutron scattering, on the other hand is dominated by deuterium scattering. Representative results for the system HCl/H_2O are shown in Figure 7.9 for different mole ratios. In dilute solution the dominant peak is due to the typical water network, with O··O distances of 285 pm. With increasing HCl concentrations a peak at 252 pm appears and is interpreted as the O··O distance corresponding to $H_3O^+ \cdot\cdot OH_2$. The main peak at 313 pm is assigned to O··Cl, with a subsidiary peak at 361 pm probably being due to Cl–Cl distances. Model calculations have suggested that the hydration structure of the proton involves four water molecules, an interpretation that is

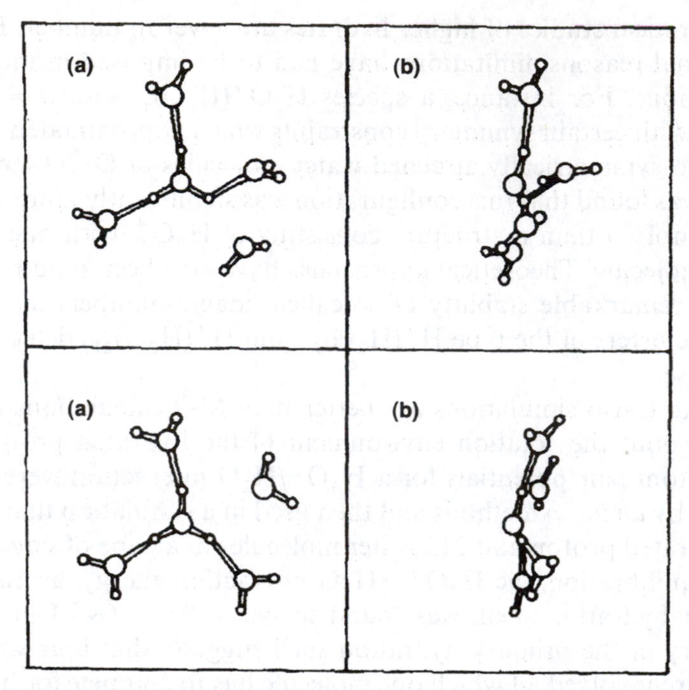

Figure 7.8 *Two typical results of a Monte Carlo simulation of $H^+(4H_2O)$ in water in top view* (a) *and side view* (b). *At any given instant only three of the four water molecules are in configurations that correspond to hydrogen bonds*
(Reproduced by permission of Cambridge University Press from D.E. Irish, *Water Sci. Rev.*, 1988, **3**, 1)

confirmed by the neutron diffraction results, although the exact position and orientation of the fourth water molecule remain uncertain.

The dynamics of proton hydration have been studied mainly by NMR and vibrational spectroscopy. In particular, interest has centred on proton-transfer processes. Whatever model is chosen, the NMR results indicate that lifetimes are in the picosecond range and that OH^- has a lifetime approximately twice that of H_3O^+. These lifetimes are several times longer than even the lowest frequency vibration of H_3O^+. If such a species exists, then it should be 'visible' in the vibrational spectrum (see below). The weakness of most NMR studies lies in the common assumption that the species involved in proton transfer is actually H_3O^+. A rate constant of the same order of magnitude would, however, be observed if the species involved was $H_5O_2^+$. Only few reports are on record where the rate of exchange has actually been measured on the water molecules.

Infrared and Raman spectroscopy of strong acids, particularly $HClO_4$, in aqueous solution have been extensively applied to investigations of the nature of hydrated proton species. Raman spectral observations on perchloric acid/water mixtures of increasing mole ratios (N) have suggested the following picture: At $N \approx 1.5$ a species H_3O^+ exists and interacts with

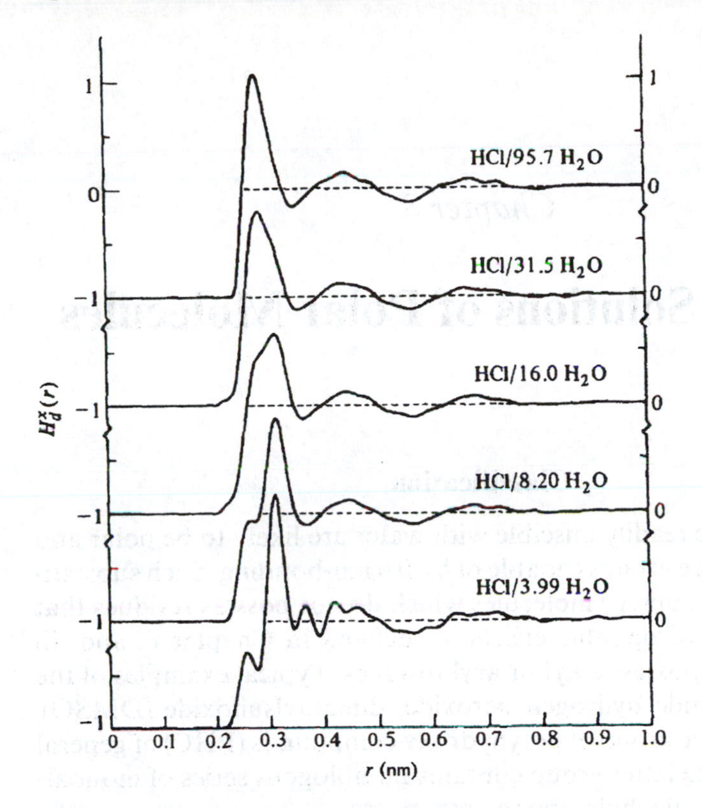

Figure 7.9 *X-ray derived atom–atom pair distribution functions of aqueous solutions of HCl at different HCl:H_2O mole ratios*
(Reproduced by permission of Cambridge University Press from D.E. Irish, *Water Sci. Rev.*, 1988, **3**, 1)

ClO_4^-. Its lifetime is sufficiently long to make it identifiable. With increasing N, the environment of the proton changes to $H_2O \cdot\cdot H_3O^+$ and the lifetime shortens significantly. Superposition of adjacent bands makes some interpretations uncertain, but spectral data from studies on aqueous glasses of HCl and HBr at liquid nitrogen temperatures have served to identify unambiguously the existence of a H_3O^+ species.

Ratcliffe and Irish conclude their definitive and exhaustive reviews of the nature of the proton in aqueous solution with the following comment: 'The "structure" adopted for the hydrated proton in water depends on the time-scale and, with or without "wishful thinking", in due time with the new picosecond technologies the flickering picture can be expected to clarify.' However, having established the (probable) existence of H_3O^+ or $H_5O_2^+$ or a mixture, hydrogen-bonded into a three-dimensional network of water molecules that even now defies a completely satisfactory description, the experimentalist still faces major unresolved problems pertaining to a very central aspect of aqueous solutions.

Chapter 8

Aqueous Solutions of Polar Molecules

Classification

Molecules that are readily miscible with water are likely to be polar and contain one or more groups capable of hydrogen-bonding. Such substances fall into two groups: (i) molecules which do not possess residues that can give rise to hydrophobic effects, as defined in Chapter 6, and (ii) molecules that do possess alkyl or aryl residues. Typical examples of the former group include hydrogen peroxide, dimethylsulfoxide (DMSO), urea and the generic group of polyhydroxy compounds (PHC) of general form $C_n(H_2O)_n$. The latter group contains homologous series of monoalkyl derivatives, e.g. alcohols, amines, ethers, etc.

Solution Behaviour and Water Structure

The above classification has already been touched upon in Chapter 6 (see Figure 6.1 and Table 6.1). Thus, solutions of polar molecules in Group 1 display a deceptively simple thermodynamic behaviour, with symmetric excess functions, and the sign of ΔG^E determined by the magnitude of ΔH^E. Deviations from ideal behaviour are marginal, an observation that has probably been responsible for past neglect of such solutions by physical chemists. Urea is the important exception; because of its ability to affect protein stability in solution, it has received considerable attention.

Early attempts to account for the solution behaviour of urea invoked a simple association model of the type

$$U_{n-1} + U = U_n \text{ etc.} \tag{8.1}$$

This model can account for the solution thermodynamics up to high urea concentrations. Its credibility came to be questioned for various reasons, but particularly because spectroscopic studies failed to provide any evidence for urea aggregates. This has also been concluded from more recent neutron scattering studies.

Equation 8.1 takes no account of the fact that water is an associated liquid, with a hydrogen-bonding topology of the tetrahedral type, whereas urea is a planar molecule with donor and acceptor sites directed in orientations which are incompatible with 'water structure' as defined in Chapter 5. Urea therefore acts as a structural perturbant, but without the necessity for specific urea–water or urea–urea interactions, such as suggested by equation 8.1, primarily because urea and water mix with an *absorption* of energy. This endothermic mixing process is, however, almost completely accounted for by the latent heat of fusion of crystalline urea. It becomes apparent, therefore, that urea removes from water the configurational property that makes water unique, namely its three-dimensional structure. Urea has been dubbed a 'statistical structure breaker'; it turns water into an 'ordinary' polar solvent. This may also account for the ability of urea to 'salt-in' noble gases, hydrocarbons, etc., albeit by a different mechanism from that caused by ions.

The type of configurational perturbation produced by urea can also be observed for hydrogen peroxide, DMSO and simple amides (glycolamide, lactamide, etc.). Unfortunately, neutron scattering has not yet been of much help in shedding light on the structural features of urea–water mixtures, because urea donates protons to water through the $-NH_2$ groups but also accepts protons through the $>C=O$ group, and the molecule lacks spherical symmetry about the nitrogen atoms. The $g(r)$ neutron data are, however, incompatible with structural features derived from computer simulation studies.

The water–water correlations in urea solutions, obtained from neutron scattering with the aid of H/D substitution, are remarkably similar to those found in $g(r,\Omega)$ for pure water (see Figure 2.4). This observation again confirms an equal probability of water–water and water–urea hydrogen bonds, i.e. water–water bonding suffers a simple 'dilution' by the presence of urea.

The situation is quite different for DMSO, a molecule that bears a superficial similarity to urea, but where hydrogen-bonding is limited to the acceptor group $>C=O$. The neutron results show that the effect of DMSO on the water–water $g(r)$ is akin to that produced by a rise in temperature.

Polyhydroxy Compounds – a Special Case

The water-like behaviour of polyhydroxy compounds (PHCs) is best illustrated by their crystal structures, which, like those of the ice polymorphs, feature infinite three-dimensional networks in which the molecules are linked in complex fashions by hydrogen bonds. That such networks, although of a more labile nature, also persist in aqueous PHC solutions, is reflected in their rheological properties. PHCs, and to an even greater extent polysaccharides, display 'interesting' rheological behaviour, which includes examples of non-Newtonian flow, gel formation, thixotropy and dilatancy.

Polymeric PHCs fulfil biological functions almost as diverse as those of proteins. Carbohydrates are also used extensively in many branches of chemical technology. Thus, 95% of all water-soluble polymers used by industry are PHC-based, mostly chemically modified in various ways. Some important biological and technological functions of PHC derivatives are summarised in Table 8.1. Despite their undoubted importance, sugars and their derivatives have long been, and still are to some extent, the Cinderellas of physical chemistry. Although crystal structures abound, systematic data on aqueous solutions are scant. The reason may well be that the measurement of PHCs' solution properties are beset by experimental difficulties, and that differences between different stereoisomers tend to be small, perhaps within the experimental errors.

It is, however, hard to see how a true understanding of PHC properties

Table 8.1 *Occurrence and applications of PHCs in biology and technology*

Function	*Example*	*Nature of PHCs*
Genetic control	Nucleotides	Ribose, deoxybribose
Energy source	Starch	Glucose (α1,4-linked)
Load-bearing fibres	Cellulose	Glucose (β1,4-linked)
Metabolic intermediates	Glycolytic products	Sugar phosphates
Sweeteners	Food additives	Sucrose, fructose
Cell–cell recognition	Blood group determinants	Mannose oligomers
Connective tissue	Hyaluronic acid	Glucuronic acid, *N*-acetyl glucosamine
Natural antifreezes	Antarctic fish antifreeze protein	Galactose
Antigens	*Bacillus dysenteriae*	Uronic acids, amino sugars
Osmoregulators	Cryoprotectants	Sorbitol, trehalose
Oil drilling fluid	Xantham gum	Glucose, mannose, glucuronic acid
Food gelling agents	Alginates	Mannuronic acid, guluronic acid

can be achieved without due consideration of the solvent contribution to chain–chain interactions, particularly, since most of the exotic rheological properties appear to be confined to *aqueous* solutions. Bearing in mind the extreme sensitivity of water structure to a variety of perturbations caused by solutes, it seems equally likely that solutes bearing water-like –OH groups, should be sensitive to the configurational requirements of the surrounding solvent medium. There now exists a wealth of experimental information confirming the sensitivity of PHC stereochemistry to solvent effects. Nevertheless, detailed studies of PHC–water interactions are still few and far between.

Compared to the much-studied hydrophobic effects, so-called hydrophilic effects are more difficult to study experimentally, more specific and more subtle. Added to the above problems is the labile nature of many PHCs, which gives rise to spontaneous intramolecular transformations, such as anomerisation, furanose–pyranose conversions, ring flexibility, etc., all of them solvent sensitive. They produce a degree of chemical heterogeneity which can cause problems in the interpretation of experimental data. This may well be the reason why, despite their ubiquity and varied functions, so little attention has been paid to the influence of the solvent, usually water, on their stereochemistry and, hence, their interactions with one another and with other types of molecules.

PHC Hydration: Theory and Measurement

Several attempts have been made to gain a more fundamental insight into the molecular details of sugar. The main features that set water apart from other liquids are its spatial and orientational intermolecular correlations, dominated by labile hydrogen-bond interactions, and its time-averaged tetrahedral geometry. PHCs contain the same chemical groups (–OH) and are able to interact with water such that, from an energetic point of view, differences between solute–water and water–water interactions are expected to be marginal, unless cooperative effects (about which little is known) act so as to favour water–water interactions. Configurationally the situation is complex because the mutual spacings and orientations of the –OH vectors in different solute molecules can match those in water to varying degrees. If a maximum degree of hydrogen-bonding is to be maintained, then in the neighbourhood of the solute the water molecules will suffer a disturbance. Bearing in mind the wide range of hydrogen-bond lengths encountered in different systems, and the ease with which the HOH angle can accommodate minor distortions, such hydration disturbances are likely to be of short-range nature.

Attention has been drawn to the spatial compatibility of the $-OH$ topology in water with that of equatorial $-OH$ groups on pyranose sugars. In the sugars the spacings between equatorially linked oxygen atoms on next-nearest carbon neighbours are 0.485 nm which is also the distance between next-nearest neighbour oxygen atoms in liquid water. Figure 8.1 illustrates how the glucose anomers could interact with an idealised, unperturbed water lattice, without causing strain in the overall hydrogen-bonding network. This hypothesis has been further developed to account for differences in the physical properties of anomeric sugar species. It runs into problems with furanoses, where the OH–OH spatial correlations on the sugar rings do not fit the water structure.

Detailed dielectric permittivity measurements on sugar solutions first revealed that all data can be fitted by two discrete Debye-type relaxation processes. One process was ascribed to the reorientation of the solvent molecules, although the relaxation time exceeded the relaxation time τ_0 of bulk water by 10%. The other process was interpreted as due to sugar-ring tumbling. The results were consistent with hydration, defined in terms of slow exchange, i.e. $\tau_{ex} > 100$ ps. Interestingly, distinct differences were observed for H_2O and D_2O solutions.

Subsequent NMR relaxation studies on ^1H, ^2H, ^{13}C and ^{17}O were performed on sugar and water molecules. Although the solvent relaxation was markedly affected by the presence of sugars, relative populations of 'hydration water' could not be resolved, because on the NMR time-scale, exchange is too fast. Remarkable differences in reorientational correlation times (τ_e) between axial and equatorial anomeric ring protons were recorded; for glucose these were 440 ps for H(1)α and 360 ps for H(1)β! The differences had already earlier been found to diminish with rising temperature. Similar differences have sometimes been interpreted in terms of the bulk viscosity, but that begs the question why ribose solutions should be less viscous than glucose solutions. The cause most probably lies in the hydration details.

According to theory, the dielectric relaxation time is related to the NMR reorientational correlation time τ_e (obtained from the nuclear magnetic relaxation rate) by $\tau_{diel} = 3\tau_c$. It has been found that for the sugars studied, this relationship is not obeyed. It appears that different relaxation processes are measured by the two techniques. Since τ_c could be shown to monitor the isotropic tumbling of the sugar ring, the dielectric measurements clearly reflect some other reorientation process, especially since τ_{diel} is found to be insensitive to the size of the sugar molecule.

All the results point to stereospecific hydration: in β-glucose, H(1) is shielded by the motion of the hydrated equatorial –OH group, with

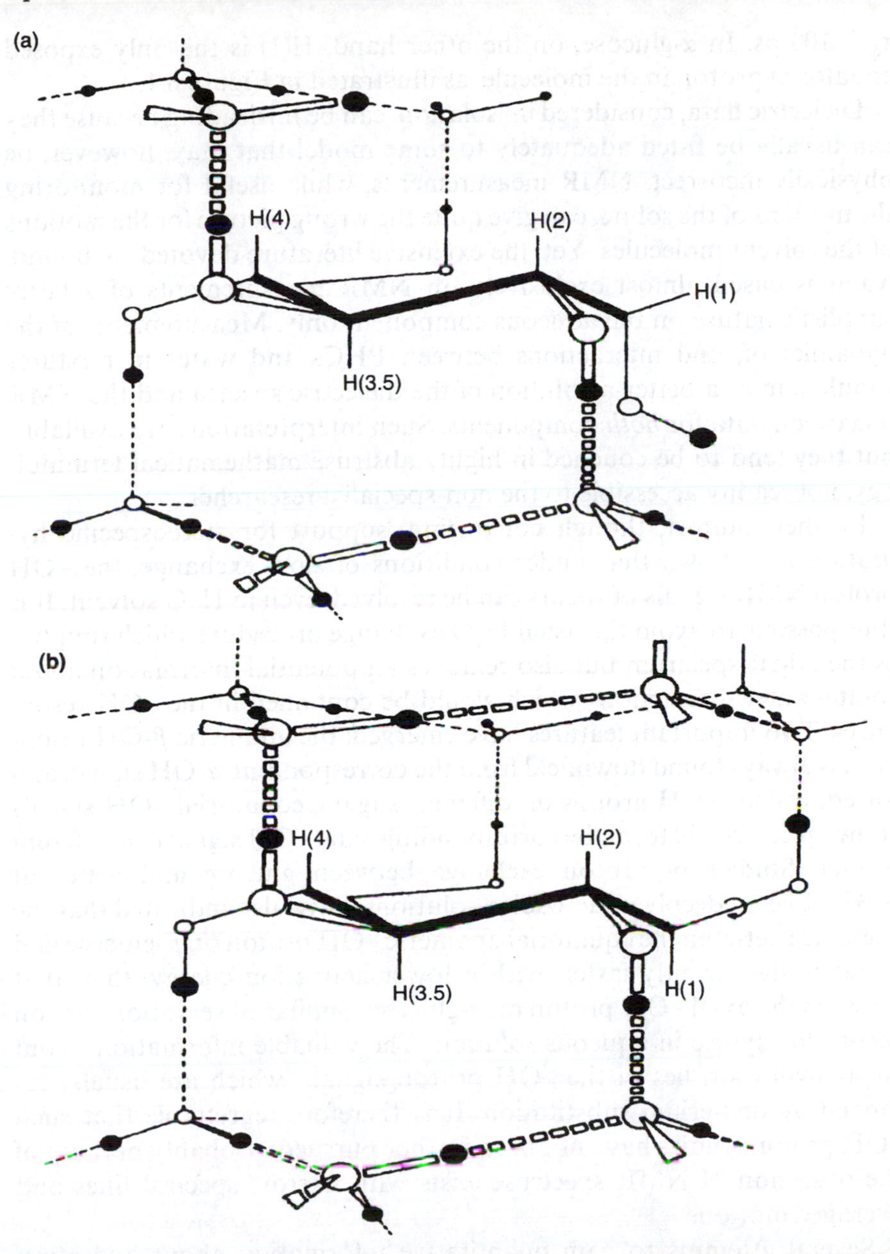

Figure 8.1 *D-glucose hydrogen-bonded into a hypothetical tetrahedral water structure, according to the specific hydration model. Water molecules below and above the plane of the sugar ring are shown:* (a) *α-anomer,* (b) *β-anomer. The pyranose ring is indicated by bold lines. Oxygen and hydrogen atoms are shown by open and filled-in circles; covalent and hydrogen bonds by solid and broken lines. The hydroxymethyl protons (H(6)) are omitted for the sake of clarity*
(Reproduced by permission of Cambridge University Press from F. Franks and J.R. Grigera, *Water Sci. Rev.*, 1990, **5**, 187)

$\tau_e < 100$ ps. In α-glucose, on the other hand, H(1) is the only exposed equatorial proton in the molecule, as illustrated in Figure 8.1.

Dielectric data, considered in isolation, can be misleading because they can usually be fitted adequately to some model that may, however, be physically incorrect. NMR measurements, while useful for monitoring the motions of the solute, can give quite the wrong picture for the motions of the solvent molecules. Yet, the extensive literature devoted to 'bound' water is based almost exclusively on NMR measurements of a fairly simplistic nature on the aqueous component only. Measurements of the dynamics of, and interactions between PHCs and water in mixtures should aim at a better resolution of the dielectric spectra and the NMR relaxation data for *both* components. Such interpretations are available, but they tend to be couched in highly abstruse mathematical terminology, not readily accessible to the non-specialist researcher.

Further indirect, though convincing, support for stereospecific hydration has shown that under conditions of slow exchange, the –OH proton NMR signals of sugars can be resolved, even in H_2O solvent. It is thus possible to avoid the usual D_2O exchange procedure which simplifies the NMR spectrum but also removes all potential information about solute–water interactions which would be contained in the –OH resonances. Two important features have emerged: the anomeric β-OH resonance is always found downfield from the corresponding α-OH signal, and for equivalent –OH groups on different sugars, equatorial –OH signals show up downfield from the corresponding axial –OH signal in a different isomer. Studies of proton exchange between glucose and water at $-35\,°C$, i.e. in deeply undercooled solutions, have also indicated that the exchange between the equatorial anomeric –OH proton on β-glucose and water is significantly faster, with a lower activation energy, than that between the axial –OH proton on α-glucose. Similar observations are on record for xylose in aqueous solution. The valuable information about sugar hydration lies in the –OH proton signals, which are usually removed by deuterium substitution. It is therefore regrettable that such –OH proton studies have not been further pursued, probably because of the obsession of NMR spectroscopists with narrow spectral lines and averaged motions.

Several attempts to gain quantitative information about hydration from near infrared or Raman spectra are on record. The physical significance of so-called hydration numbers is questionable, because they are obtained from the proportion of the water signal intensity that cannot be detected, ostensibly because 'bound' water does not contribute to the observed intensity. This type of reasoning is also quite frequently applied in spectroscopic studies of water in biological systems.

The value of thermodynamic studies has already been discussed in Chapter 6. The available vapour pressure, volumetric and calorimetric data provide some surprising information about PHC hydration and interactions between PHC molecules in solution. Inspection of the thermodynamic excess functions (equation 6.7) for the two stereoisomeric monosaccharides glucose and mannose, shown in Figure 8.2, reveals a behaviour that is typical for all PHCs: deviations from ideal solution behaviour, as reflected by $G^E(m)$, are marginal, even up to high concentrations. That the mixtures are, however, far from ideal is shown by the non-zero nature of H^E. The lack of ideality is hidden by the almost complete compensation of H^E by TS^E, yet another unique feature of aqueous solutions.

Early vapour-pressure measurements on aqueous solutions of some common sugars led to the assumption that simple hydration models could accommodate deviations from ideal solution behaviour. Hydration was expressed in terms of 'hydration numbers' n_h, i.e. the number of bound water molecules per sugar molecule. The thermodynamic properties of aqueous solutions of a sugar S could then be expressed by a series of simple equilibria of the type

$$S_{i-1} + H_2O \rightleftharpoons S_i \qquad i = 1,2, \ldots, n_h \tag{8.2}$$

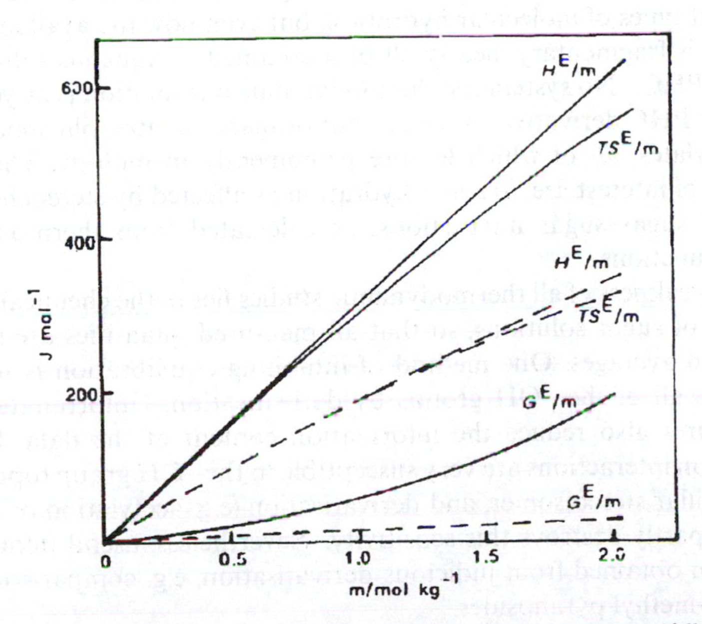

Figure 8.2 *Excess thermodynamic quantities, at 25 °C, of aqueous glucose (full lines) and mannose (broken lines) solutions*
(Reproduced by permission of Cambridge University Press from F. Franks and J. R. Grigera, *Water Sci. Rev.*, 1990, 5, 187)

Assuming all hydration sites (and therefore all equilibrium constants K_{i-1}) to be equivalent, an average hydration number could be derived which depended only on the water chemical potential (water activity). Remarkably, such a simple model adequately accounts for the water activity of sugar solutions at 25 °C up to high concentrations. For instance, glucose obeys equation 8.2 up to saturation ($K_n = 0.789$, $n_h = 6$) and sucrose up to 6 M ($K_n = 0.994$, $n_h = 11$). The weakness of such simple hydration models lies in their incorrect prediction of the sign of the enthalpy of dilution. The prediction was that, with increasing temperature, the equilibrium would be driven to the right, a result that is inconsistent with experiment and with physical reality. As a rough-and-ready correlation, equation (8.2) may have its uses for the prediction of some technological or biological attributes of complex, carbohydrate-containing systems. As a credible physical model, the concept of simple hydration, based only on the number of –OH groups, should be discarded. It cannot account for the properties of aqueous solutions. The pretence that the solution properties of PHCs can be so simply accounted for has delayed more penetrating studies into their physical chemistry for many years.

The first systematic thermodynamic studies on simple sugars in solution date from the 1970s when attempts were made to characterise different types of molecular hydration, but even now the available information is fragmentary, nearly all of it confined to aqueous solutions of simple PHCs. No systematic thermodynamic information is as yet available for PHC derivatives, such as aminosugars, sulfates, phosphates and carboxylates, all of which feature prominently in biology. The major aspects of interest are: (i) sugar hydration as affected by stereochemistry, and (ii) sugar–sugar interactions, as calculated from thermodynamic excess functions.

The weakness of all thermodynamic studies lies in the chemical heterogeneity of sugar solutions, so that all measured quantities are suitably weighted averages. One method of inhibiting equilibration is to block some or all of the –OH groups by derivatisation. Unfortunately such procedures also reduce the information content of the data, because hydration interactions are very susceptible to the –OH group topology of a particular stereoisomer, and derivatisation (e.g. acetylation or esterification) partly destroys this sensitivity. Nevertheless, useful information has been obtained from judicious derivatisation, e.g. comparisons of α- and β-1-methyl pyranosides.

As regards useful information about hydration details, the limiting partial molar heat capacity C^0 is probably the most sensitive thermodynamic indicator. C^0 is here defined by

$$\Delta C^0 = C^0 - C^* \qquad (8.3)$$

where C^* is the heat capacity in some reference state, usually taken to be the crystalline state. C^0 is obtained experimentally either from direct heat-capacity measurements or from the temperature dependence of the heat of solution of the solid PHC.

Heat-capacity measurements have perforce to be made on the equilibrium mixture which, in the case of sugars, may consist of several isomeric species. For heat of solution measurements, the starting material will normally be a pure crystalline substance. On the assumption that its conversion to the equilibrium mixture is slow, then the measured value refers to a solution of the pure isomer. In order to compare these data with those obtained from equilibrated solutions, it is necessary to correct for the ΔC which accompanies the equilibration. Thus, for the mutarotation of glucose, starting from the α-anomer, $\Delta C = -9$ J mol^{-1} K^{-1}. The anomeric mixture contains 37% of the α-anomer, and therefore C^0 of the equilibrium mixture should be increased by 5 J mol^{-1} K^{-1} for comparisons with data obtained by heat of solution measurements on the pure α-form.

C^0 measures a combination of the effects of hydration (solute–solvent interactions) and any changes in the internal degrees of freedom of the solute molecule when it is removed from its crystalline environment. Thus, C^0 is a measure of the effect produced by a change in environment on the number of degrees of freedom, mainly intermolecular, of the system. As expected, C^0 is determined mainly by molecular size and complexity. Nevertheless, second-order effects are apparent, e.g. in the C^0 values of the three pentitols listed below. In their crystalline forms, ribitol (**A**) and xylitol (**B**) exist with bent carbon chains (sickle-shape), whereas arabinitol (**C**) has a planar zig-zag carbon atom arrangement. It might be expected that in solution, ribitol and xylitol should show some similarities in their thermodynamic behaviour, but differ from arabinitol. This is, however, not the case: xylitol has a C^0 that is significantly lower than those of the other two isomers. A similar difference is observed for the hexitols, where sorbitol (glucitol) has a lower C^0 than its isomers, galactitol and mannitol. Large positive C^0 values are usually taken as an indicator of thermolabile hydration spheres. On that basis it appears that the hydration structures surrounding xylitol and sorbitol are of a shorter range and/or less complex than those associated with their isomers, a conclusion that is consistent with MD simulation results, described below.

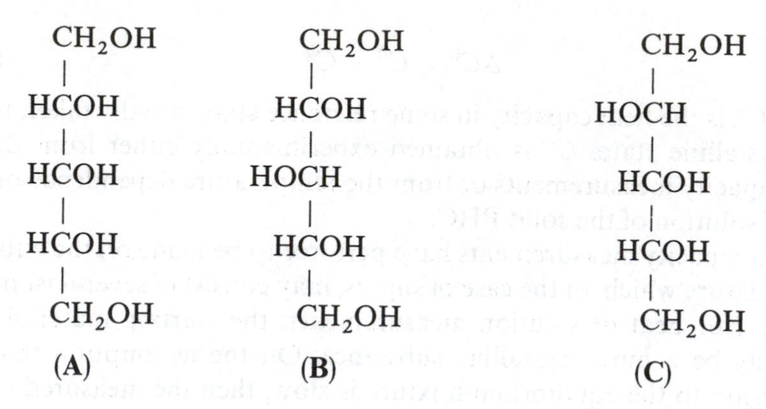

Equation 8.3 can also be written for the volumetric properties, i.e. for the evaluation of the limiting partial molar volume V^0, provided that the necessary V* data are available. The limiting thermodynamic quantities of some PHCs are collected in Table 8.2. V^0 is also mainly a function of the molecular weight, with each additional (CH_2O) contributing approximately 16–20 cm^3 mol^{-1}. A general observation is that V^0 values of PHCs are surprisingly small compared to their non-hydroxylated analogues. Thus, the introduction of a CH_2 group into the tetrahydrofuran ring, to form tetrahydropyran, increases V^0 by 15 cm^3mol^{-1} which is almost the same as the increment produced by the addition of –CHOH into the furanose ring, despite the considerable difference in the group molecular weights. The comparison of furan and pyran derivatives with furanoses and pyranoses suggests that the density measurement does not 'see' the –OH groups as part of the solute molecule but as water molecules. Observations of this type raise questions whether the V^*-values, calculated from van der Waals radii, are realistic reference states, and if not whether the V^0 data so obtained can usefully be compared with those referred to crystal densities.

Influence of Water on Solute–Solute Interactions

Interest here centres on the effect of stereochemical detail on molecular pair interaction coefficients in equation 6.8, extracted from the measured thermodynamic excess functions. Table 8.3 contains selected data for the free energy and enthalpy. While results from calorimetric studies abound, there is little information about g_{ii}. The reason is clear: vapour-pressure measurements on dilute solutions are difficult to perform with the necessary degree of precision, especially where deviations from ideal behaviour are marginal. By comparison, enthalpy determinations are more straightforward, because of the availability of high-precision calorimeters. The

Table 8.2 *Limiting partial molar volumes (V°) and heat capacities (C°) of PHCs in aqueous solution at 25 °C*

PHC	V° ($cm^3 mol^{-1}$)	V^* ($cm^3 mol^{-1}$)	C° ($J\ mol^{-1}\ K^{-1}$)	C^* ($J\ mol^{-1}\ K^{-1}$)
β-Arabinose	93.7	92.4	278(D), 270 (L)	184
Xylose	95.4	98.5	279	184
Ribose	95.2	94.4	276	187
β-Fructopyranose	110.7	112.7	338	232
α-Galactose	110.5	111.3	352	217
β-Galactose			324	216
α-Mannopyranose	111.5	115.2	337	220
α-Glucose		115.3	347	224
α/β-Glucose	112.0	118.3	331	
β-Glucose		116.1	346	
α-Me glucoside	133.6		435	
α-Me galactoside	132.6		436	
β-Me galactoside	132.9		445	
α-Me mannoside	132.9			
Trehalose	206.9			
Cellobiose	213.6			
Maltose	208.8		614.2	
Sucrose	211.3		649.4	
Lactose	207.6		619	
Cellotriose	309.2			
Maltotriose	304.8			
Raffinose	302.2		931	
Ethane diol			193	
Glycerol			240	
Erythritol			310	
Ribitol			376	
Xylitol			346	
Arabinitol			374	
Mannitol			452	
Glucitol (sorbitol)			412	
Galactitol			445	

V^* and C^* refer to the crystalline reference state.

scattered nature of the data precludes the establishment of patterns in the relationships between the stereochemistry and the g_{ii} coefficients. It was thought earlier that g_{ii} was positive for all PHCs, except *myo*-inositol which is based on the cyclohexane ring and has a high degree of internal symmetry. However, both galactose and mannose appear to have $g_{ii} < 0$.

As regards the available h_{ii} data, mixing appears to be endothermic, except for some hexitols where exothermic mixing has been observed. There is no obvious relationship between molecular size or number of

Table 8.3 *Free energy and enthalpic molecular pair interaction parameters, see equation (6.8)*

PHC	g_{ii}[a]	h_{ii}[a]
Xylose	81.5	332 (D), 336 (L)
Ribose	7.5	202
Arabinose	34	196 (D), 192 (L)
Glucose	63	342
Galactose	65	133
Mannose	43	207
Fructose	83	264
2-Deoxyribose	80	464
2-Deoxyglucose	88	592
2-Deoxygalactose	58	442
1-Me-α-xyloside	114	1126
1-Me-β-xyloside	100	1098
1-Me-α-galactoside	33	900
1-Me-β-galactoside	134	1081
1-Me-α-glucoside	171	1097
1-Me-β-glucoside	82	1048
1-Me-α-mannoside	144	1206
Cellobiose	138	764
Maltose	100	483
Trehalose	140	595
Lactrose	195	506
Sucrose	205	577
Raffinose	353	458
Myo-inositol	−260	−800

Units: J kg mol^{-2}.

–OH groups and h_{ii}, but chemical modification (deoxysugars, methyl pyranosides) leads to increases in h_{ii}. As was the case for the limiting thermodynamic properties, measurable differences in h_{ii} exist for diastereoisomers, e.g. ribitol, xylitol and arabinitol, emphasising once again that sugar–sugar interactions in solution are sensitive to the –OH group topology. With the present state of our knowledge, no quantitative interpretation of the data can be attempted. Also, the numbers presented in Table 8.3 must still be treated with a degree of caution, since few of the data have been confirmed by several independent groups of workers.

With a few exceptions, the general rule seems to be that for PHCs, $h_{ii} > g_{ii} > 0$ which contrasts with urea-like compounds for which $h_{ii} < g_{ii} < 0$. It has been suggested that the observed behaviour can be attributed to a partial reversal of hydration. Such a process is speculated to involve only minor changes in the energy and degrees of freedom of the water released from the 'more structured hydration cospheres of the saccharides' to the bulk, with increasing concentration.

Several attempts are on record for the prediction of thermodynamic properties on the basis of group additivity schemes. The particular property in question is treated as the linear sum of contributions from the chemical groups that make up the molecule, e.g. $>CH_2$, $>CHOH$, –O–, –OH, etc. Although such empirical schemes have been successful in accounting for the properties of homologous series of hydrocarbons and monofunctional alkyl derivatives, they clearly cannot discriminate between stereoisomers and cannot therefore provide a reliable method for the prediction of PHC properties in aqueous solution.

Hydration and Computer Simulation

It is now clear that direct studies of PHC hydration are beset by many problems, arising mainly from the fact that the lifetimes of hydration 'structures' in solution are generally shorter than the time-scales of most experimental methods. The incomplete realisation of this problem has led to much confusion and many misunderstandings regarding the existence and nature of so-called 'bound' water. Those who interpret experimental observations in terms of bound water seldom define such binding, whether in terms of $g(r)$ (distances), binding energies or lifetimes. Computer simulation, provided the necessary care is taken in the formulation of potential functions, can be of help in defining hydration geometries and exchange rates, at least to a first approximation.

Molecular dynamics results on glucose in water demonstrate the effect of water on the conformation and dynamics of the system. Although the average sugar-ring conformation is not much affected by hydration, the solvent acts so as to increase the amplitudes of structural fluctuations. Changes are observed in the $-CH_2OH$ conformation as a result of hydration. The radial distribution functions of water in the hydration sphere (nearest-neighbour molecules) show up clear differences between –OH and $-CH_2OH$ hydration structures. Thus, apolar hydration is found in the neighbourhood of the $>CH_2$ group, with 12.3 nearest-neighbour water molecules at a distance of < 0.49 nm. In contrast, the –OH groups have only 3.2 nearest water neighbours, with adjacent –OH groups sharing water neighbours by hydrogen-bonding, a structure which may be significant for anomerisation. Characteristically, the hydration of the ring oxygen differs from that of the exocyclic oxygen atoms. As expected, water molecules in the oxygen hydration shells are also orientationally restricted. The simulation provides clear evidence for intramolecular hydrogen-bonding that is preserved (most of the time) even in aqueous solution! This finding is in direct contrast to the experimental NMR results (in D_2O solution). If true, such bonds would probably impose very definite –OH group orientations, with correlated –OH rotations. However, the simulation results suggest that –OH rotations are uncorrelated, but slower than *in vacuo*.

Confusing descriptions, like (water) structure making and breaking, found their way into the literature during the 1960s and gained widespread popularity; PHCs generally are classified as structure makers. The MD simulations on glucose clearly demonstrate that, despite the well-defined and localised hydrogen-bonding between sugar –OH groups and water, all nearest-neighbour water molecules exchange rapidly with bulk water. Water can therefore be regarded as 'bound' in a structural sense, but not in a dynamic sense, since exchange with 'bulk' water takes place on a <10 ps time-scale.

MD simulations on mannitol and sorbitol, not being subject to anomeric and tautomeric complexities, might provide more exact information on PHC interactions with water, albeit 'computer water'. In terms of lifetimes, hydration can be defined by a parameter $R = \tau_h/\tau_0$, where the τ values describe the average lifetimes of a water molecule in the hydration shell (h) and in bulk (0) respectively. Simulation results show that both polyols have $R < 1$, corresponding to *negative hydration*. However, the R-values differ for the two diastereoisomers: 0.80 for mannitol and 0.39 for sorbitol, highlighting once again the extreme sensitivity of water to the stereochemical detail of the PHC molecule.

Since dynamic data cannot provide structural information, it is unwise to assign structure making/breaking properties. What is evident, however, is that both hexitols (as well as glucose) generate mobile water environments in their proximities. Hydration numbers can be assigned, being the average number of water molecules at a distance of 1.2 molecular diameters from the hexitol molecule. This yields $n_h = 13.2$ and 11.5 for mannitol and sorbitol, respectively. The ratio, 1.55, is close to the experimental ratio of their calculated hydrodynamic volumes, 1.28.

In this connection, another MD study of sorbitol and mannitol is of interest. Considering that these two molecules only differ in the position of the –OH group on C(4), one might expect similar conformations in solution. For the *in vacuo* simulation, this is indeed found to be the case. No difference can be observed in the end-to-end distances or the radii of gyration (0.54 nm) of the two molecules. However, in aqueous solution ('computer water'), the radii of gyration decrease to 0.417 nm for mannitol and 0.372 nm for sorbitol, irrespective of the starting (crystal) configuration, i.e. planar zig-zig or bent carbon chain. These values are peculiar to the aqueous solvent and differ markedly from those obtained with a Lennard–Jones solvent or an argon-like solvent in which the Lennard–Jones parameters are chosen to match those of the water oxygen atom.

Solvation and PHC Conformation

Early ^{13}C and ^{1}H NMR studies on alditols in solution led to the claim that their molecular configurations were identical to those found in the crystals. This implies that (i) distinguishable conformations do exist in solution but (ii) that such conformations are insensitive to solvent effects. More recent studies demonstrated that ^{13}C NMR spectra, taken on their own, can lead to erroneous conclusions, and also that potential-energy calculations, based on the summation of group interactions derived from studies of six-membered rings, are not necessarily applicable to acyclic molecules.

The ambiguities of physico-chemical measurements on unmodified sugars result largely from the unavoidable heterogeneity of their solutions. On the other hand, chemical modifications of sugars, which can prevent anomerisation and/or tautomerisation, may also obscure or distort the very properties which are of interest. A useful way of studying the effects of stereochemical detail on solvation, conformation and weak solute–solute interactions is with the aid of acyclic PHCs, such as the sugar alcohols, the crystal structures of which are known. Experimental results can then be related unambiguously to unique stereochemical species.

Presumably, the transfer of a molecule, say a pentitol, from the crystal to an aqueous solution takes place with the substitution of ten solute–solute hydrogen bonds by ten (?) solute–water bonds. It is not obvious why this transfer should not affect the stability of the particular molecular configuration as it exists in the crystal. For obvious reasons, ^{1}H spectra of PHCs, alditols in particular, are extremely complex. Even at 400 MHz, the ^{1}H spectra required extensive computer simulation and deuterium labelling. Higher field-strength (620 MHz) spectra of pentitols and hexitols in D_2O and pyridine-d_6 could, however be resolved and all signals unambiguously assigned (see Figure 8.3). The measured J_{HH} coupling constants suggest rapidly interconverting mixtures of *trans* and *gauche* conformers. With the aid of crystallographic data, the fractions of each conformer in the *trans*-configuration have been calculated. In the high-temperature limit, one might expect free rotation about all bonds. The departure from this 'free-rotation' limit for three stereoisomers is shown in Table 8.4.

Two effects can be distinguished: in pyridine the alditols have a higher degree of rotational freedom than in D_2O presumably due to decreased

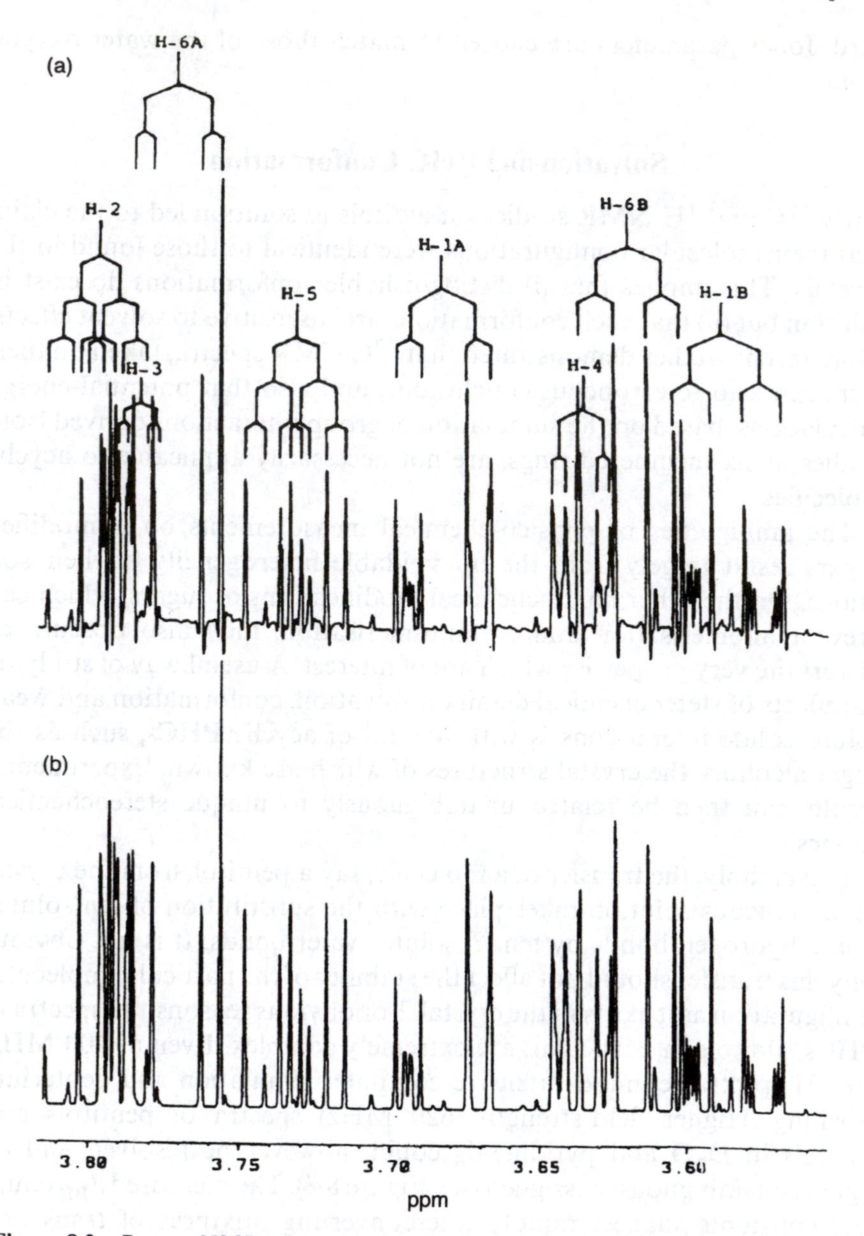

Figure 8.3 *Proton NMR spectrum of sorbitol in D_2O:* (a) *experimental spectrum with signal assignments and* (b) *computer simulated spectrum, based on the assignments*
(Reproduced by permission of Cambridge University Press from F. Franks and J.R. Grigera, *Water Sci. Rev.*, 1990, **5**, 187)

Table 8.4 *Percentage departure from free rotation; for explanation see text*

Solvent	*Ribitol* (%)	*Xylitol* (%)	*Arabinitol* (%)
D_2O	28	17	30
Pyridine-d_5	19	12	23

solvation effects. Of greater importance is the observation that in D_2O xylitol shows significantly freer rotation than either of the other two isomers. Whereas the crystal structures suggest a similarity in conformation between ribitol and xylitol (both have sickle-shaped carbon chains), thermodynamic solution properties (Tables 8.2 and 8.3) suggest a similarity between arabinitol and ribitol but a unique behaviour for xylitol. The results *in toto* therefore support the view that the measured thermodynamic properties reflect the conformation in solution and are not related to the crystal structure.

It is hardly surprising that, in view of the many water-like properties of PHCs, the solvent should have an impact on the structural and dynamic behaviour of these molecules in solution. Table 8.5 summarises the compositions of some anomeric equilibrium mixtures in a variety of solvents. In the simplest cases, e.g. glucose, the equilibrium mixture only contains the two anomeric forms of one of the possible ring conformers; in this case it is the Cl pyranoid ring. In many other cases, several tautomers and their anomeric forms coexist. The equilibrium mixture of ribose contains at least six distinct species: the anomeric forms of each of the 1C and Cl pyranoses and of the furanose ring. The composition of the equilibrium mixture depends on the solvent. Potentially the situation is somewhat more complex than is shown in Table 8.5, because several pyranose conformers might coexist. Furthermore, the sugar concentration also affects the equilibrium composition, as well as its temperature dependence.

All aldohexoses (with the exception of altrose) which exhibit complex mutarotation, i.e. where more than one tautomer is involved, have an axial –OH on C(4). This is also the case for those pentapyranoses which exhibit complex mutarotation. These observations have led to the generalisation that an axial –OH group on C(4) of a pyranose leads to appreciable proportions of furanose forms, although no reasons have been advanced for this 'rule'. The mutarotation of β-fructopyranose in water does not produce the corresponding α-anomer but a mixture of α- and β-fructofuranose. This has been rationalised with the aid of a comparison of the favoured $^5C_2$ conformations of the α-pyranose forms of allulose, tagatose and sorbose which do not have an axial –OH at C(5) and

Table 8.5 *Tautomeric and anomeric equilibrium percentage compositions of sugars in solution*

Sugar	*Temperature (°C)*	*Solvent*[a]	*Pyranose*		*Furanose*		
			α	β	α	β	
Ribose	30	D	24	52	9	15	
	30	DMSO	16	55	6	23	
Xylose	31	D	36.5	63		<1	
	25	P	45	53	1	1	
Galactose	31	D	30	64	2.5	3.5	
	25	P	33	48	7	12	
Glucose	30	D	34	66			
	30	DMSO	44	56			
Mannose	44	D	65.5	34.5	0.6	0.3	
	116	P	78	22			
	116	DMSO	86	14			
Fructose	30	D	2	70	5	23	
	80	D	2	53	10	32	Keto: 3
	33	P	5	43	15	35	
	33	DMSO	5	26	21	48	
Sorbose	27	D	98		2		
	80		87	2	8	1	
Ribose-5-P	6	H			35.6	63.9	Keto: 0.1
Arabinose-5-P	6	H			57.3	40.4	Keto: 0.2
Xylose-5-P	6	H			70.5	24.8	Keto: 0.3
Fructose-6-P	6	H			16.1	81.8	Keto: 2.2
Fructose-1,6-P_2	6	H			13.1	86.0	Keto: 0.9

[a] H = H_2O, D = D_2O, P = pyridine-d_5, DMSO = dimethylsulphoxide-d_6.

β-fructose which does have an axial –OH(5) group in its favoured tautomer, the $^2C_5$ pyranose.

In aqueous solution the pyranose ring of reducing sugars appears to be more stable than the furanose ring, although this situation is reversed for some sugar derivatives. In non-aqueous solvents appreciable proportions of furanoses exist in the equilibrium mixtures; the reason is not immediately obvious but must reside in the solvation contributions to the conformational free energies of the various species and the ease of interconversion. Accepting then, that water stabilises the pyranose ring, the next question concerns the relative stabilities of the two anomers. Despite several efforts to calculate conformational free energies, it is still not obvious why the $\alpha:\beta$ ratio for glucose is 36:64, whereas for mannose it is 67:33. Conventionally, such observations have been rationalised (with the benefit of hindsight) by invoking a specific, but empirical 'anomeric effect'.

Alternative explanations of anomeric equilibria take explicit account of solvation as a determining factor. Given that the intrinsic conformational free-energy differences between coexisting species, as calculated from vacuum-based potential functions, are only of the order of kT, it is deduced that the solvation contribution plays an important but so far unexplained role in determining the equilibrium composition.

Of all chemical species, the PHCs most closely resemble water in their properties. The purpose of the above, somewhat lengthy discussion of PHC interactions, hydration and conformation was to demonstrate how much remains to be done to bring some order into a research area that is beset by experimental complexities. It should, nevertheless, be of fundamental scientific interest and is undoubtedly of great technological and biological importance (see also Chapter 13).

Hydrophobic/Hydrophilic Competition

Many important classes of water-soluble molecules are constructed from chemical groups, the combination of which results in a competition between hydrophobic hydration and direct hydrogen-bonding effects. They include homologous alkyl and aryl derivatives such as alcohols, amines, amides, ketones as well as more complex molecules, such as peptides, surfactants, phospholipids and a variety of polymers, e.g. polyvinyl alcohol, polyethylene glycol, polyvinyl pyrrolidone and hydroxyalkyl cellulose and starch derivatives. All these substances display complex solution behaviour.

Alcohol–water mixtures have received the most intensive study and in the following paragraphs their solution properties are summarised as being typical of most other water-soluble organic molecules which contain apolar residues but are also able to interact with polar solvents, and sometimes with one another by direct hydrogen-bonding.

Until recently, all structural evidence was of a circumstantial nature, some of it already referred to in Chapter 6; see, in particular, Figure 6.1. Alcohol–water mixtures, as a generic group, display positive deviations from ideal solution behaviour ($\Delta G^{E} > 0$), which originate from configurational constraints, i.e. $T\Delta S^{E} << 0$. The observation that $\Delta H^{E} < 0$ was rationalised by invoking alcohol–water hydrogen-bonding. A multitude of scattering, NMR and spectroscopic studies and computer simulations, performed over the past 50 years, have confirmed the dual 'schizophrenic' hydrophobic/hydrophilic solution behaviour, with marked anomalies at low concentrations. These anomalies show themselves as a non-monotonic concentration dependence of many measured properties (see Figure 6.4). Positions of the minima and/or maxima depend on the temperature and the hydrophobe/hydrophil group ratio (HLB values to the surface chemist). It is instructive to compare the four related compounds, all derivatives of *n*-propane and also their isomers:

CH_3	CH_3	CH_3	CH_2OH
|	|	|	|
CH_2	CH_2	CHOH	CHOH
|	|	|	|
CH_3	CH_2OH	CH_2OH	CH_2OH

The magnitudes of $T\Delta S^E$ decrease with an increase in the number of –OH substitutions, with glycerol–water mixtures behaving almost ideally. It also appears that 'compact' molecules, such as *t*-butanol, behave more ideally than their straight-chain isomers, i.e. *n*-butanol.

The maxima/minima in many physical properties suggest a fundamental structural change at concentrations which, again, depend on the hydrophobe/hydrophil group ratio. The results are consistent with a picture of water cage formation being dominant in dilute solution. With increasing concentration, the quasiclathrate structure becomes unstable and is replaced by a more normal mixture in which limited alcohol aggregation (embryonic micellisation) can be observed by light and small-angle X-ray scattering methods, as shown in Figure 8.4. At even higher concentrations the peculiar behaviour is replaced by a more normal mixing.

In the past, quantities such as $g(r,\Omega)$ and $u(r)$ could only be inferred, although the available results strongly pointed towards the interpretation described above. There is now limited neutron scattering information against which to test earlier speculations. The available experimental data are still limited, but use of the H/D substitution method has produced the first direct structural information which can be summarised as follows:

(1) In a 10 mol % solution of methanol, the water environment of the $-CH_3$ group resembles the clathrate cage structure, with its preference for O–H vectors oriented tangentially along the sides of the polyhedral cage. As might be expected, all cages are distorted and there is no evidence of long-lived hydration structures.

(2) The $g_{HH}(r)$ functions for solutions of ethanol and *t*-butanol are identical, i.e. the water molecules do not discriminate between the two alcohols. Unfortunately the alcoholic –OH group contribution to $g_{HH}(r)$ has not been, and cannot be, isolated. At the concentration used, 3 mol % *t*-BuOH, it is considered to be negligible. The superposition of the $g_{HH}(r)$ functions onto that of pure water reveals that only very minor changes on 'water structure' are produced by the presence of the alcohol.

The competition between hydrophobic and direct hydration (hydrogen-bonding) effects plays a basic role in biology and many branches of

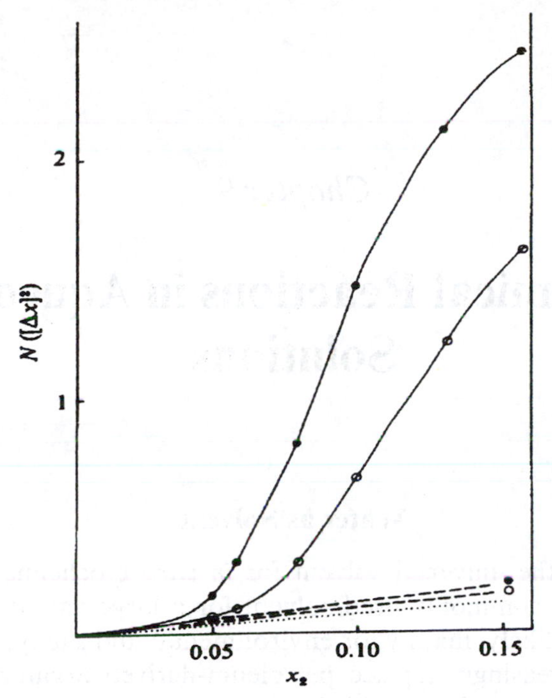

Figure 8.4 *Concentration and temperature dependence of the mean square concentration fluctuations on tert-BuOH–water (solid line) and dioxan–water (broken line) mixtures at 17.5 (open circles) and 49.5 °C (solid circles), derived from light scattering intensities. The dotted line indicates the expected fluctuations for an ideal solution*
(Reproduced by permission of Cambridge University Press from F. Franks and J.R. Grigera, *Water Sci. Rev.*, 1990, **5**, 187)

technology. Processes such as emulsification, micellisation, phase behaviour of water-soluble polymers and lipids, rheological properties of concentrated solutions, etc., all depend critically on this phenomenon which is peculiar to water as solvent. It originates from the unique nature of the water molecule, with two proton donor and two proton acceptor sites, directed towards the vertices of a tetrahedron. Thus, water 'prefers' to hydrogen-bond to itself (cf. ice). Failing that possibility, the next best course is to form hydrogen-bonds to other hydroxylic molecules, e.g. PHCs. Where even that is impossible, water molecules will perform orientational reorganisations that make possible the maintenance of three-dimensional hydrogen-bonded water networks, with the concomitant formation of cages. This configuration is capable of accommodating nonpolar guest molecules, albeit a price must be paid in terms of the permitted orientational degrees of freedom. This is the origin of the hydrophobic effect, the biochemical ramifications of which will be further explored in Chapter 10.

Chapter 9

Chemical Reactions in Aqueous Solutions

Water as Solvent

Besides being the universal solvent for *in vitro* biochemical processes, water is also a commonly used solvent for a large variety of chemical reactions. It is likely, mainly for environmental and safety reasons, that water will increasingly replace petroleum-derived organic solvents in process chemistry. Although this is considered to be generally desirable, several problems may arise that set water apart from other solvents. Aqueous solubilities of many organic materials are severely limited, making it necessary, for economic reasons, to employ mixed aqueous solvents in large-scale manufacturing processes. This will reduce the benefits of 'clean' water over 'dirty' petrochemicals and necessitate the provision of additional waste-disposal stages into chemical processing.

Surprisingly, there is still a lack of basic engineering data, such as those required to develop efficient and cost-effective processes, even for well-known reactions. Some years ago, the author set a class of chemical engineering students the task of developing an efficient process for the production of salicylic acid by the Kolbe reaction between phenol and carbon dioxide, i.e. calculating optimum concentrations, temperatures and pressures. It turned out that neither the thermodynamic nor the kinetic parameters governing this simple reaction are recorded in the literature. Apparently the current process was developed empirically.

In earlier chapters it was stressed that, in principle, aqueous systems can be adequately treated by the various theories now available to physicists and chemists. In practice, however, this is often impossible, because several unique features of aqueous systems are likely to invalidate some simplifying assumptions that form the basis of currently available theories. These features, some of which have already been discussed in Chapters 6–8, arise from the unique geometry of the water molecule and its ability to participate in three-dimensional structures, stabilised by weak forces. The introduction of solute species modifies these forces and their directionality. It is therefore not surprising that reaction rates and mechanisms in aqueous media are sensitive to such effects, which tend to be particularly pronounced in mixed aqueous solvents. In practice, aqueous reaction media are nearly always complex mixtures, and in some cases they produce some remarkable solvent effects on reaction rates and associated quantities.

Water Participation in Chemical Reactions

The involvement of water in physico-chemical or biochemical processes is difficult to quantify in molecular terms, so that observed solvent effects are usually accounted for in terms of some macroscopic property, such as viscosity or dielectric permittivity. The same is true for solvent effects on reaction rates, but such devices are of little help where a detailed reaction mechanism is to be elucidated. In fact, solvent effects partly arise from the *participation* of the solvent in the reaction sequence, for instance in a proton donor/acceptor role. Alternatively, solvent involvement in a reaction sequence can be by oxidation/reduction mechanisms, furnishing electrons which are used in some other step of the sequence. Both the acid/base dissociation of water and its oxidation/reduction potential can be sensitively dependent on the presence of co-solvent and it is probably this sensitivity which is responsible for some remarkable solvents (see below).

A common, simplifying assumption is usually employed, when water is replaced by an aqueous/organic solvent mixture, namely that the addition of the organic component only alters the bulk properties of the reaction medium, e.g. melting point, viscosity, dielectric permittivity, but does not affect the kinetic characteristics of a chemical transformation that is performed in this mixed solvent medium. Possibly such an assumption is valid for some simple reactions, but it has been shown to be invalid for biochemical processes. For instance, the luciferase-catalysed oxidation of reduced flavine mononucleotide, which proceeds by a bioluminescent pathway in water, gives rise to the same end product in water–ethanediol mixtures, but by a different, dark pathway.

The Self-ionisation of Water

The self-ionisation of water probably lies at the very basis of aqueous solvent participation effects. At its simplest, the equilibrium is expressed by

$$H_2O\,(liq) \rightleftharpoons H^+(aq) + OH^-(aq) \qquad (9.1)$$

By (aq) we mean hydrogen and hydroxide ions, as they exist in the solution under consideration, irrespective of the solvation details of these species, which were described in Chapter 7. The equilibrium constant K_w is commonly referred to the hypothetical unit molality standard state for solutes (*i.e.* the two ions) and the pure liquid for the solvent. For dilute solutions:

$$K_w = a_{H^+} a_{OH^-}$$

The corresponding standard free energy and its various derivatives are then readily obtained from K_w measurements at several temperatures and pressures, or preferably by direct calorimetric and dilatometric measurements. Table 9.1 summarises the 'best' available data for the dissociation of water. It is seen that pK_w is very temperature sensitive. For instance, at $-35\,°C$, in undercooled water, $pK_w \sim 17$, a result which must have fundamental consequences in low-temperature adaptation mechanisms of living organisms. At high temperatures and pressures K_w exhibits a complex behaviour, but the generalisation can be made that water becomes increasingly more ionised. For instance, at ordinary temperatures one water molecule in 5.5×10^8 is ionised, but at 1000 °C and 10 MPa this has increased to one in 840. Extrapolation of presently available results suggests that at still higher temperatures water will be almost completely ionised and that its properties may then resemble those of fused hydroxides.

Although experiments with D_2O have not been so numerous, it appears that at 25 °C, H_2O is more ionised than D_2O, and also that the

Table 9.1 *'Best' standard thermodynamic functions for the water ionisation equilibrium at 298.2 K*

$pK_w = 13.9991$
$\Delta G^0 = 79.830$ kJ mol^{-1}
$\Delta H^0 = 55.76$ kJ mol^{-1}
$\Delta S^0 = 80.71$ J (mol K)$^{-1}$
$\Delta C_p^0 = 223.6$ J (mol K)$^{-1}$
$d(\Delta C_p^0)/dT = 3.8$ J (mol K^2)$^{-1}$
$\Delta V^0 = -22.12 \times 10^{-6}$ m^3 mol^{-1}

difference in ΔH^0 contributes more than does $T\Delta S$ of ionisation. Care must be taken in such comparisons with the correct definition of standard states, because of the differences in molecular weight and density between H_2O and D_2O.

Turning now to the ionisation of water in two- or multicomponent systems, the most systematic studies are those performed on electrolyte solutions with the aid of emf measurements on electrochemical cells of the type

$$H_2 \mid HX,MX \mid AgX, Ag$$

and

$$H_2 \mid MOH,MX \mid AgX, Ag$$

By extrapolation to zero ionic strength (pure water) it is possible to obtain the activity coefficients of H^+ and OH^- at each electrolyte concentration.

In the presence of organic additives one would expect K_w to increase, and this has indeed been found, except for ethanediol. Reported pK_w values are based on the assumption that the organic solvent does not contribute to the hydrogen-ion activity. This is of course only justified in the case of the ideal inert co-solvent, but not when the co-solvent itself possesses acid/base properties. For reasonably strong acids (acetic acid) or bases (aniline), an analysis of the emf data is straightforward, but for aqueous mixtures of, say, methanol, glycerol or glucose the experiments and their interpretations are laborious, but they can yield very accurate ionisation constants for the organic components.

Any reasonable understanding of acids and bases in aqueous solution (and that includes most molecules) must be very much concerned with the interactions of the various ions and molecules with water. Since H^+ and OH^- are ever-present, they will be taken as typical examples. The most reliable data come from gas-phase mass spectroscopic measurements which have enabled equilibrium constants for the following equilibria to be established:

$$H^+(H_2O)_n(g) + H_2O(g) \rightleftharpoons H^+(H_2O)_{n+1}(g)$$

and

$$OH^-(H_2O)_n(g) + H_2O(g) \rightleftharpoons OH^-(H_2O)_{n+1}(g)$$

The corresponding ΔH and ΔS values have been obtained from measurements at several temperatures and are summarised in Figure 9.1 as a function of *n*.

The quantities of real interest are the *hydration energies* of *single ions*, i.e.

$$H^+(g) \rightarrow H^+(aq)$$

but these can only be obtained by the combination of thermodynamic and spectroscopic measurements with extrathermodynamic assumptions based on various theoretical calculations. Estimates for ΔG^0 range from -982 to -1220 kJ mol^{-1} and, taking a mean value, it is then possible to calculate $\Delta H^0 = -268$ kJ mol^{-1} for the equilibrium

$$H^+(H_2O)_8(g) \rightleftharpoons H^+(aq)$$

which suggests that in the complete hydration of a proton many water molecules are involved. In other words, the representation of the hydrated proton in solution by the species H_3O^+ or even $H_9O_4^+$ is not particularly realistic (but refer to Chapter 7).

Ionisation Reactions

The thermodynamics of acid/base ionisation have been a popular subject for study over many years. Ingenious calorimetric methods have been

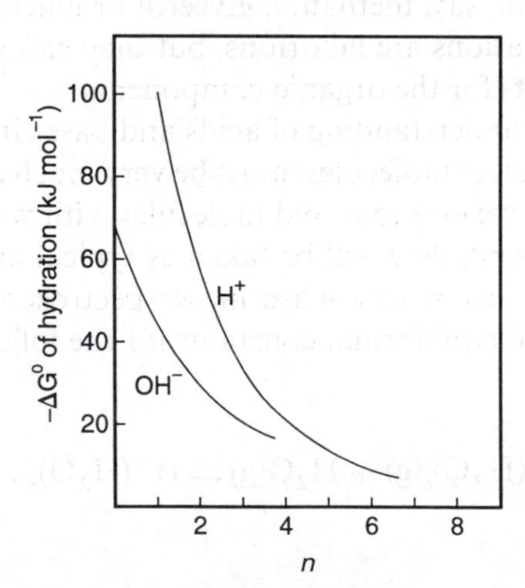

Figure 9.1 *Hydration free energies of H^+ and OH^- in the gas phase, as functions of the number of water molecules in the hydration shell*

devised that allow for the simultaneous measurement of pK, ΔH^0 and ΔS^0. Substituent effects on the ionisation of related organic acids clearly point to the involvement of the solvent in such processes. Such effects can be quantified by comparing the ionisation of some reference acid HR^z with that of another acid HA^z. For proton transfer between the acids, we have

$$HA^z(S) + R^{z-1}(S) \rightleftharpoons A^{z-1}(S) + HR^z(S) \tag{9.2}$$

where S is the solvent medium. Thermodynamic quantities for the equilibrium of equation (9.2) are obtained as $K_s = K_{HA}/K_{HR}$, and

$$\delta(\Delta X^0) = \Delta X^0{}_{HA} - \Delta X^0{}_{HR}$$

where the left-hand side represents the substituent effect on some thermodynamic property X. A widely used, and very successful empirical correlation of substituent effects is the Hammett equation

$$\log K_s = \rho\sigma \tag{9.3}$$

where σ depends only on the substituent and the parameter ρ depends on the type of reaction, the solvent and the temperature. The temperature insensitivity of σ provides for other useful relationships. As a first approximation $\delta(\Delta H^0)$ and $\delta(\Delta S^0)$ can be assumed to be invariant with temperature (i.e. a zero heat capacity change), when

$$\delta(\Delta H^0) = \beta\delta(\Delta S^0) \tag{9.4}$$

where β is an integration constant that is identical with the so-called isoequilibrium or isokinetic temperature. When $\delta(\Delta C^0) \neq 0$, this relationship takes on a slightly more complex form. Equation 9.3 has found extensive use, but it has become well known that it often provides a good correlation to equilibrium constants and free energies even when enthalpies and entropies vary in an apparently erratic fashion. Nowhere is this more common than when the solvent is an aqueous/organic mixture. The question is then why the Hammett equation works better than it should. Or to rephrase the question: how is it possible to ignore solvent and temperature and to explain substituent effects on K_s solely in terms of potential energies of solute species? This apparent paradox has been resolved by expressing both $\delta(\Delta H^0)$ and $\delta(\Delta S^0)$ as sums of 'internal' and 'environmental' contributions. For many symmetric reactions $\delta(\Delta S_{int}) \approx 0$, so that $\delta(\Delta S^0) \approx \delta(\Delta S_{int})$. It has also been found that there is generally an almost complete compensation between $\delta(\Delta H_{env})$ and $T\delta(\Delta S_{env})$. The result is that chemical substitution effects are quite well accounted for by the 'internal' enthalpy (potential energy). The Hammett equation thus gives a good approximate correlation of equilibrium constants and free energies even when enthalpies and entropies are not in accord with the thermodynamic requirements of the equation. This is one further example of the statement made earlier that free energies generally disguise more than they reveal.

The phenomenon of enthalpy–entropy compensation has been subjected to detailed study; it appears to be particularly pronounced for aqueous solutions. When water or some mixed aqueous solvent forms the environment in which the reaction takes place, then it stands to reason that $\delta(\Delta H_{env})$ and $T\delta(\Delta S_{env})$ can be identified with the structural (or solvation) changes that take place in the medium. It seems then, that such changes provide an almost zero contribution to the free energy of the process under study, despite the fact that $\delta(\Delta H_{env})$ and $T\delta(\Delta S_{env})$ can be very large and of either sign. Therefore, if the aim of a series of experiments is to study the role of the solvent in a given process, equilibrium constant measurements at a single temperature are unlikely to provide useful information. Rather, experiments should be performed at several temperatures (and pressures, if possible) and the free energy derivatives obtained which contain more significant information about solvent effects.

There are, however, also aqueous solvent mixtures where the Hammett equation breaks down and where K_s is not a linear function of solvent composition. This then raises the question of changes in the actual nature of the 'solvated proton' in solution. Thus, in mixtures of water and methanol, K_a shows a non-monotonic solvent composition dependence, with a well-formed minimum on the dilute solution side. This implies that the reference acid, used to study the process shown in equation 9.2, changes its nature in a complex manner with composition.

Despite all the various pitfalls, the chemical literature abounds in K_s data for related acids and bases, performed at a single temperature, where authors have taken liberties in rationalising apparently anomalous results in terms of '*specific*' *solvation interactions*, solvent structure, changes in the degree of hydrogen-bonding within the solvent, differences in the hydration of molecules and ions, etc. *Caveat lector*.

Chemical reactions in which water is one of the reactants (or products) are basically of four types: oxidation, reduction, hydrolysis and condensation. For obvious reasons these reactions play a crucial role in life processes. They are therefore discussed in more detail in Chapter 11.

Reactions in Mixed Aqueous Solvents

The physical organic chemist is well aware of the bizarre solvent effects observed in mixed organic/aqueous media. Principally, three approaches have been used to account for such behaviour. Before discussing them, let us examine some examples of aqueous solvent effects on reaction kinetics. Figure 9.2(a) shows the rate constant (normalised to that for pure water) for the water catalysed detritiation of *t*-butylmalononitrile in water–DMSO mixtures. One would expect that the addition of the less-polar solvent DMSO ($\varepsilon = 6$) would reduce the reaction rate, but that is only observed for solutions quite concentrated in DMSO. Interestingly, the maximum in the curve corresponds to a maximum in the kinetic basicity of the solvent mixture, but it is not immediately clear how this would produce the observed rate behaviour. Extrema in rate constants are very rare; a much more common type of behaviour is shown in Figure 9.2(b) which represents the kinetics of the water catalysed hydrolysis of *p*-methoxyphenyl dichloroacetate in aqueous mixtures of *t*-BuOH. Here $-\Delta G^{\ddagger}$ ($-RT\ln k$) increases monotonically with solvent composition, while at the same time, $\Delta H^{\ddagger}$ and $T\Delta S^{\ddagger}$ exhibit a very complex behaviour which is, nevertheless, very common for reactions in mixed aqueous solvents where the organic component is a hydrophobic-type molecule and is present in concentrations of up to 15 mol %.

In attempts to rationalise mixed solvent effects the first approach has

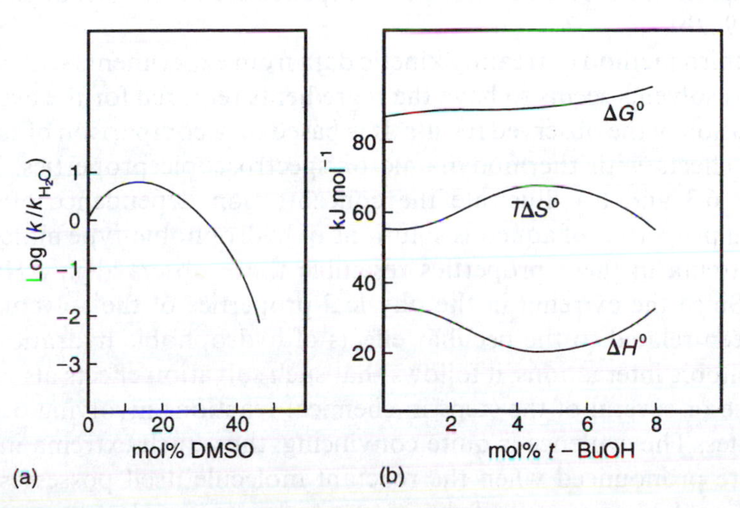

Figure 9.2 *Effects of mixed solvents on the kinetics of organic reactions:* (a) *the detritiation of* t-*butylmalanonitrile,* (b) *the water catalysed hydrolysis of* p-*methoxyphenyl dichloroacetate*

been to relate rate constants to some macroscopic property of the solvent. Various functions of ε have been used with limited degrees of success. Other parameters that have been tried include empirical polarity functions based on charge transfer spectra, dipole moments, polarisability, viscosity, internal pressure, nucleophilicity and ionising power. While it is sometimes possible to correlate rate constants with one or other of these functions, we must once more remember that free energies can be easily fitted by most properties. None of the above parameters can account for the enthalpies of activation observed in mixed aqueous solvents with high water contents. One is driven to the conclusion that the structural integrity of the solvent is an important factor in influencing the rate-determining step, usually proton transfer between the solvent and the substrate.

A second approach to account for solvent effects relies on the concept of preferential solvation. According to this, one or the other of the solvent species in the mixture (H_2O and S) preferentially solvates the reactant molecule X. Depending on the mole ratio of the two solvent species, solvation can then be expressed in terms of a series of equilibria of the type:

$$X.H_2O_n \rightleftharpoons X.H_2O_{n-1}S \rightleftharpoons X.H_2O_{n-2}S_2 \ldots$$

Such preferentially solvated states have indeed been shown to exist, particularly in ionic solutions, and they can account for non-linearity in reaction rates with changes in solvent composition. They cannot, however, explain the solvent composition dependence of $\Delta H^\ddagger$, as depicted in Figure 9.2(b).

The third method of treating kinetic data from experiments with mixed aqueous solvents seems to have the ingredients required for the eventual explanation of the observed results. It is based on a comparison of solvent kinetic effects with thermodynamic or spectroscopic properties. Thus, Figures 6.3 and 6.4 illustrate the concentration dependence of some physical properties of aqueous solutions of hydrophobic-type molecules. The extrema in these properties resemble those observed in $\Delta H^\ddagger$ and $T\Delta S^\ddagger$. Since the extrema in the physical properties of the solvent itself have been related to the peculiar effects of hydrophobic hydration and hydrophobic interactions, it follows that such solvation effects also influence one or several of the steps in chemical reactions involving organic substrates. The evidence is quite convincing: thus, such extrema in $\Delta H^\ddagger$ are more pronounced when the reactant molecule itself possesses substantial apolar residues, and they move to lower co-solvent concentrations but increase in amplitude as the alkyl group in the co-solvent molecule is increased in size.

A useful treatment has been devised to examine whether the co-solvent effect operates mainly through changes of solvation of the reactant ground (initial) state or through the transition state, since $\Delta H^{\ddagger}$, as measured experimentally, would be the sum of both such effects. Similar dissections of experimental data have been performed for the volume of activation, $\Delta V^{\ddagger}$, where it seems that mainly the transition state is affected.

Mixed solvent effects are particularly marked in the heat capacity of activation, $\Delta C^{\ddagger}$. For instance, for the solvolysis of *tert*-butyl chloride in an ethanol–water mixture of ethanol mole fraction x, $\Delta C^{\ddagger}$ changes from -320 at $x = 0$ to -800 at $x = 0.05$ and to -60 J mol^{-1} K^{-1} at $x = 0.1$. The exact significance of these large changes is not clear, but they mirror similar heat capacity phenomena observed in the effect of ethanol on the $N \rightleftharpoons D$ equilibrium of proteins in solution (see Chapter 10).

In summary, it appears that any molecule or ion that can perturb the structure of water is likely to affect the activation parameters of reactions in such mixed aqueous media. The most dramatic results are observed not in $\Delta G^{\ddagger}$ itself but in its pressure and temperature derivatives, even when $\Delta G^{\ddagger}$ itself varies in a monotonic fashion with solvent composition. In order to interpret kinetic data in mixed solvents, it is therefore necessary to ascertain how the intermolecular nature of water itself is modified by the presence of the co-solvent, to give rise to a completely 'new' solvent, rather than a simple mixture of water molecules and co-solvent molecules.

Chapter 10

Hydration and the Molecules of Life

Water Structure as a Determinant of Biological Function

Ignorance still surrounds the origin of life's molecular building blocks. It has been shown that a reducing atmosphere of methane, ammonia and water vapour, when subjected to electric discharges, can generate amino acids. On the other hand, the high energies required for the synthesis of even primitive proteins could also have been obtained in superheated water under the influence of solar radiation at a time when such radiation could reach the earth unhindered by a shielding magnetic field. Rudimentary forms of life could thus have developed in volcanic cracks and crevices at the temperature of molten lava.

Irrespective of the origin of nature's molecular and supermolecular building materials, a given biological function and activity is generally associated with a highly specific three-dimensional structure, maintained largely by weak, noncovalent forces, the formation and stability of which require the involvement of water. In the case of proteins, folding takes place on the ribosome in an aqueous medium of complex composition.

In earlier chapters it was demonstrated that the so-called 'liquid water structure' is labile and extremely sensitive to dissolved molecular species of all types but that, in turn, water also affects the intra- and intermolecular behaviour of such dissolved molecules. It is therefore hardly surprising that those molecules, which form the building blocks of all living matter, should also be sensitive to the nature of their aqueous environment. This chapter provides a summary of evidence for solvent involvement in the generation of so-called native structures of biopolymer molecules and their assembly to supermolecular structures, such as multisubunit enzymes, membranes, phages and viruses.

The molecules in question are proteins, nucleotides, lipids and oligo- and polysaccharides. Details of the solvent involvement in promoting biologically significant structures are not very well established, owing to the complexity of the molecules involved. There can, however, be no question that hydration phenomena do play a dominant role in the promotion of 'native' structures, structures that are associated with life processes. This whole subject could itself provide enough material for a book, so that, within the confines of a single chapter, only a few salient issues can be discussed. Emphasis will be placed on proteins in solution. Because of their enormous structural and functional diversity their hydration behaviour has received most attention by researchers.

Chapter 8 developed the theme that quantitative structural, energetic and dynamic descriptions become less precise and more speculative as systems increase in complexity. Hard information about the role of hydration in the formation of biologically active structures, as derived from *in vitro* experimental studies, will therefore remain speculative for some time to come. On the other hand, in computer simulation and modelling investigations, the assumptions and simplifications that must of necessity be applied render the value of such approaches questionable, except when performed by the most knowledgeable and sceptical practitioners.

Of the four groups of molecules, lipids are distinct from the other three categories because they are 'small' molecules, rather than polymers, so that the discussion will centre on intermolecular processes. With biopolymers, on the other hand, much of the interest lies in intramolecular, conformational phenomena.

Comparison of the Molecules of Life

The essential architectural differences between the three biopolymer groups are outlined in Table 10.1. Proteins have identical backbone structures and in a sense they resemble synthetic homopolymers. They derive their chemical variety from the amino acid composition and the sequence of the amino acid residues along the peptide chain. Differences between carbohydrates, on the other hand, arise mainly from different

Table 10.1 *Architectural comparison of different biopolymer types*

Feature	*Proteins*	*Carbohydrates*	*Nucleotides*
Monomer variety	20+	Usually 1–3	4
Linkage	Identical	Variable	Identical
Chain branching	Never	Common	Never
Chemical modification	Seldom	Common	None

types of linkages between chemically similar sugar residues, and from chain branching. Thus, cellulose and amylose are chemically identical, both being homopolymers of glucose, but they are physically very distinct. The former is a β1,4-linked polyglucose, and the latter has α1,4 glucosidic linkages. Nucleotides, like proteins, have identical chemical linkages without chain branching. They can, like polysaccharides, exist as cyclic polymers, e.g. plasmids. In any one nucleotide chain, the number of residues is limited to four bases (two purines and two pyrimidines), variety being achieved by the sequence distribution. In a double-stranded nucleotide, variety is produced by the interchain hydrogen-bonded links between pairs of residues.

The term 'chemical modification' in Table 10.1 serves to describe the abundance of naturally occurring derivatives. Modified amino acids in proteins are a rarity and are usually associated with undesirable (pathological) changes in biological function and/or activity. By contrast, polysaccharide derivatives are common in biomolecules. Modifying groups which can replace –OH include $-NH_2$, –COOH, SO_4^{2-} and others. Such modifications can render the resulting polymers sensitive to pH and solvent medium changes.

The chemical variety displayed by lipids is extensive. Molecules of biological importance are mainly derived from glycerol through mono, di or trisubstitution. They can be polar, e.g. phospholipids, or nonpolar, e.g. glycerides or sugar lipids. The alkyl chains exhibit a wide variety of lengths and degrees of unsaturation. A common feature of lipids is their amphiphilic character, which gives rise to different types of aggregation structures, e.g. micelles, bilayer membranes and vesicles. In the formation of all such structures the solvent medium plays a dominant role.

A study of water involvement in the formation and stabilisation of bioactive structures must address the following questions:

(1) Where are the functional water molecules with respect to the surface of the biomolecule, what is the geometrical water structure near the surface and how far into the bulk does the perturbation extend?
(2) To what extent are water molecular degrees of freedom coupled to those of the biomolecule and what are the time-scales for water exchange?
(3) How are the above properties linked to biological function; could such emerging knowledge be utilised in 'protein engineering'?
(4) How precise does such knowledge have to be for it to assist usefully in computer simulations of equilibrium and kinetic rate processes involving biomolecules?

Protein Stability *In Vitro*

The phenomenon of solvent involvement in affecting biopolymer stability in solution has been studied most intensively in proteins. At the outset it must be remembered that the word 'protein' itself describes a mixture of a polypeptide chain *and* water, and that in aqueous media the polypeptide chain can exist in a unique conformation that renders it biologically active, where it is normally referred to as 'native'. The stability of native proteins *in vitro* is severely limited by environmental conditions, such as pressure, temperature, pH and solvent composition.

Developments in X-ray and, particularly, neutron diffraction methods have led to significant progress in the elucidation of water-molecule coordinates in small peptide and protein crystals. It must, however, be remembered that, unlike the amino acid residues that make up the covalently linked peptide chain, water molecules interact with the chain only by hydrogen-bonding and are labile, subject to more or less rapid exchange. Even in crystal diffraction studies, therefore, one is not observing actual water electron densities or neutron intensities, but *probability densities*, with lifetimes governed by exchange rates. The situation in liquid or *in vivo* environments is even more complicated. Despite a vast, and rapidly growing literature devoted to protein folding, there is as yet little real understanding about the molecular and energetic details of hydration interactions and their essential role in determining the conformational or functional attributes of proteins. Incorporation of such factors into theoretical and computer simulation procedures presents severe challenges, but protein folding and stability results, arrived at without attempts at including hydration effects, must at best be misleading and of questionable value. Direct protein hydration studies should be based on dynamic methods of measurement, usually NMR relaxation. Such measurements are informative, because they provide lifetimes and exchange rates. They also require a high level of expertise and are laborious to perform.

During the past 50 years, the subject of protein stability has been subjected to an ever-increasing scrutiny by biochemists and, latterly, also by molecular biologists, biotechnologists, physicists and theoreticians. Despite all the concentrated activity, some of the very central questions still remain to be answered. It is, in fact, questionable whether on a cost/understanding basis the effort has been particularly productive. We have certainly witnessed the reinvention of the wheel several times over, and also the introduction of some remarkable hypotheses, almost bordering on metaphysics and alchemy.

Since interest in proteins is, rightly, centred on their behaviour in aqueous media, the following discussion aims to summarise current thinking and outstanding questions. The simplest model treats protein stability in terms of two states which coexist in dynamic equilibrium:

$$\underset{\text{native state}}{\mathrm{N}} \rightleftharpoons \underset{\text{denatured state}}{\mathrm{D}}$$

The N-state can be uniquely defined in terms of structural specificity and biological activity. The 'denatured' state presents more problems. It has often been equated with the flexible 'random coil', as defined in polymer science. This is, however, an oversimplification, because the rigidity of the C–N bond does not allow for complete flexibility. Inactivation, usually known as denaturation, can be associated with changes in one or more levels of structure. The most sensitive monitor of denaturation, however, is a change in, or the disappearance of, biological (or immunological) activity. The description of the D-state is therefore somewhat vague; it is now known that the nature of one or several coexisting D-states depends on the manner in which the protein is denatured.

According to the simple two-state N/D model, denaturation and renaturation are cooperative processes, so that there exist no intermediate species with appreciable lifetimes or concentrations. Under properly controlled experimental conditions, denaturation and renaturation have been shown to be reversible, and they can therefore be treated by simple equilibrium thermodynamics and the process characterised by an equilibrium constant K:

$$K = \frac{[D]}{[N]} \tag{10.1}$$

According to the conventions of equilibrium thermodynamics

$$\Delta G^0 = -RT\ln K \tag{10.2}$$

and

$$\frac{\mathrm{d}\ln K}{\mathrm{d}T} = \frac{\Delta H}{RT^2} \tag{10.3}$$

and

$$\Delta H = \Delta C \mathrm{d}T \tag{10.4}$$

where ΔC is the isobaric heat capacity change accompanying denaturation. This quantity may itself be a function of T, e.g. $\Delta C = A + BT + DT^{2+} \ldots$

While the simple two-state model may not in its strictest sense be applicable to many proteins, the introduction of possible intermediate states into the above equations does not, in principle, alter its basic tenets.

One of several surprising results to have emerged from the many studies of the denaturation/renaturation process is that, even under the most favourable conditions, the conformational stability (free energy) margins of native proteins hardly ever exceed 50 kJ mol^{-1}, corresponding to the energy of no more than three hydrogen-bonds. It is obvious, therefore, that whatever may be the stabilising influences, they are balanced almost completely by destabilising effects, leaving only a marginal net free energy of stabilisation

The physical properties of water are sensitive to the same factors that influence protein stability, so that some connection is likely. Probably the three types of hydration discussed in earlier chapters play the major role. Accordingly, amino acids are grouped into three categories: ionogenic, apolar and polar. It is now recognised that, in order to maintain a stable globular structure, a peptide chain must contain at least ca. 50% of apolar residues. These residues also tend to be more highly conserved than the polar residues. In the N-state they form the structural core of globular proteins, whereas the polar and ionogenic residues tend to be located on the periphery or flexible loops and are associated with the biochemical function of the particular molecule.

The group of globin proteins exhibits an extreme example of the above statement: despite their functional diversity, their three-dimensional structures are identical, and of the 54 apolar residues that they contain, 44 are completely conserved, whereas none of the 21 polar amino acids is fully conserved. This illustrates the general principle that buried residues tend to be more fully conserved than solvent-exposed residues. It also raises doubts about the principle of structure/function relationships, so beloved by life scientists.

Detailed studies of protein crystal structures led to the concept of 'water accessibility' of amino acid residues in the folded and unfolded states. A spherical water molecule of a given radius is allowed to roll over the surface of the protein molecule and the degrees to which the individual residues are accessible to the probe molecule are expressed by an accessibility index, which ranges from zero for a buried residue to unity for a fully solvent-exposed residue. Changes in solvent accessibility during unfolding are then believed to be related to stability and ΔG^0 is expressed in terms of the accessibility, summed over all residues. This type of approach is very sensitive to the chosen water-molecule radius and it also assumes, although implicitly, that water is a close-packed liquid. It further assumes that the native protein molecule is a stationary structure, not subject to conformational fluctuations, which would alter the temporary degree of exposure of certain residues to the solvent. Water accessibility estimates are thus seen to be of doubtful value as methods of probing water involvement in stabilising or destabilising unique conformational states.

Following on from equations (10.1)–(10.4), Figure 10.1 illustrates a typical protein thermal stability profile, together with the associated thermodynamic functions describing the reversible unfolding/refolding processes. Two important points are apparent: (i) the strongly curved $\Delta G(T)$ profile is indicative of a large specific heat change, and (ii) the small free energy change, even under optimum conditions, results from an almost complete cancellation of enthalpic and entropic contributions, neither of which needs to be small. This latter effect is again one of the mysteries of aqueous solutions, referred to as enthalpy/entropy compensation, already mentioned in the previous chapter.

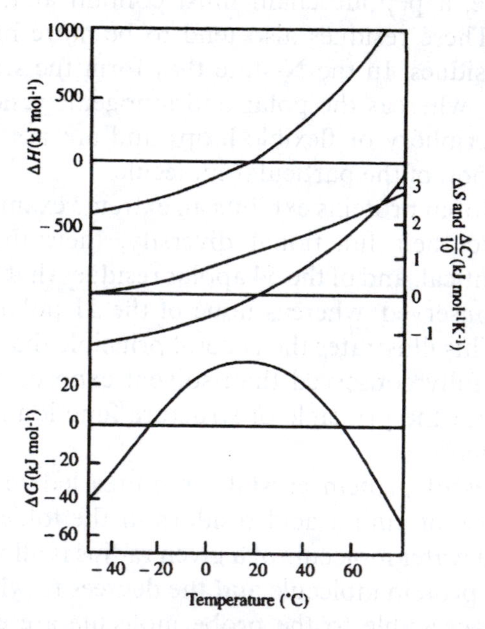

Figure 10.1 *Thermodynamic functions describing the stability profile of lactate dehydrogenase from measurements in the neighbourhood of* T_L *and* T_H *and curve fitting; from top to bottom:* ΔH, $\Delta C/10$, ΔS, ΔG
(Reproduced by permission of Academic Press from F. Franks, *Adv. Protein Chem.*, 1995, **46**, 105)

Heat and Cold Inactivation

Examination of the lactate dehydrogenase results in Figure 10.1, which are typical of proteins in general, also reveals that, even under optimum pH conditions, the stability of the native state is limited to the relatively narrow temperature range $-20 < T\,°\mathrm{C} < 55$. In other words, a uniquely ordered, folded structure can be destabilised by heating *or* by cooling. This is almost tantamount to saying that ice can be melted either by heating or by cooling! The phenomenon of cold inactivation clearly indicates that the delicate stability balance of proteins is perturbed in distinctly different ways at the two temperature extremes. These differences have as their common basis the temperature sensitivity of the physical properties of the common solvent: water. The major contributors to $\Delta G(T)$ are probably apolar hydration/aggregation effects, intrapeptide effects, configurational entropy and hydrophilic/ionic hydration. Of these, the first two are of a general stabilising nature, whereas the latter two favour the unfolding of the polypeptide chain. Table 10.2 summarises the major interactions and the (probable) signs of their temperature coefficients. Thus, interactions for which $\mathrm{d}(\Delta G)/\mathrm{d}T > 0$ weaken at low temperature, and vice versa. The net effect of temperature changes is then to perturb the delicate balance between large stabilising and destabilising contributions which, under physiological conditions, is responsible for the marginal stability of active proteins.

At the molecular level the causes of high- and low-temperature inactivation are seen to be quite different. Probably the main cause of cold inactivation is the weakening of the collective water–apolar group repulsions, which provide the main driving force for maintaining the folded

Table 10.2 *Major contributing interactions to* in vitro *protein stability and temperature coefficients of contributing interactions*

Interaction	*Contribution to* $(\Delta G)_{N\to D}$	$\left[\frac{\mathrm{d}(\Delta G)}{\mathrm{d}T}\right]_{N\to D}$
Hydrophobic interactions	+ + +	+
Salt bridges	+ +	+
Configurational free energy	− −	−
Intrapeptide H-bonds	+ +	−
Water-peptide H-bonds	− −	−
Van der Waals	+	−
Polyelectrolyte effect	−	?

+ = positive contribution to stability, − = negative contribution (favouring denatured state), ? = unknown contribution. The larger the number of symbols the greater the contribution.

structure under *in vitro* conditions at 'normal' temperatures. A subsidiary driving force for cold inactivation is the increasing affinity of ionic and polar groups for water. In the language of polymer science, water becomes a 'good' solvent (by direct water–solute hydrogen-bonding) at low temperatures. The temperature dependence of the heat capacity of cold denaturation lends support to these conclusions.

Cold inactivation of proteins is quite distinct from freeze inactivation (see Chapter 13) and must be studied in homogeneous solution, i.e. in the absence of ice. Most reported investigations have involved partial destabilisation by factors other than temperatures, e.g. pH, so-called cryosolvents or mutants. The information so obtained is, at best, ambiguous. Thus, comparative studies of enzyme-catalysed processes have shown that reactions carried out in cryosolvents (aqueous solutions containing high concentrations of freezing-point depressants, e.g. DMSO or methanol) proceed along pathways that differ from those associated with *in vivo* conditions. On the other hand, these same reactions, performed *in vitro* in undercooled water, proceed along the physiological pathways. In order to obtain definitive results, therefore, temperature should be used as the sole means of perturbation. In most cases this requires the use of undercooled water. Although experimentally inconvenient, the accessible temperature range can then be extended down to $-40\,°C$ (the practical homogeneous nucleation limit of ice; see Chapter 12).

A particular advantage of combining heat and cold denaturation in studies of protein stability is that it provides two experimental windows of $\Delta G(T)$, albeit each is limited to $0 \pm 5\ \text{kJ mol}^{-1}$, over which the stability profile can be probed, without any necessity for other changes in the environmental conditions.

One important result to have emerged from direct studies of cold denaturation concerns the heat capacities of the N and D states,. Based on results from heat denaturation studies, it had for many years been held that ΔC accompanying the inactivation of proteins is positive, large and independent of temperature. In Figure 10.1 it is convincingly shown that ΔC is in fact a non-linear function of temperature and would, in principle, be expected to change sign at some low temperature. The implications regarding the different mechanisms of the inactivation processes at the two temperature extremes are obvious.

An unresolved mystery remains: according to the two-state equilibrium model described by equation 10.1 etc., ΔG^0 and the derived thermodynamic properties are continuous functions of temperature and the D-states obtained by denaturation at the two temperature limits must then be identical. This is, however, hardly borne out by experiment; quite distinct differences between the two D-states can be observed, a result that is hardly surprising in the light of the data in Table 10.2.

Finally, still on the subject of proteins and hydration, the mechanism of water transport across biological membranes has for long been a mystery. Until recently it was believed that such transport was of the passive type, relying on leaky phospholipid domains. The discovery of specific water-transporting protein channels, now named aquaporins, has provided a new insight into water transport across cell membranes. They are of major importance in the process of water purification by the kidneys, and other organs. Several multimeric aquaporin structures have now been identified and are currently being subjected to detailed studies.

Hydration of 'Dry' Proteins

If a protein solution is carefully concentrated under conditions that do not disrupt the N-state (i.e. moderate temperature), then in some favourable cases, water can be almost completely removed and an almost anhydrous protein obtained, although this may appear to be contradictory to the very definition of a protein. If water vapour is then readmitted, the interactions in the system can be monitored by any one of several physical or biochemical techniques, and the sites of interactions characterised as a function of the water content. Table 10.3 summarises the techniques and the type of information that have been obtained.

A popular method for studying water-sensitive or water-soluble polymers (and this includes proteins) is by the measurement of water vapour sorption isotherms. Usually the sorption and desorption isotherms do not coincide. Sorption hysteresis at low water contents is symptomatic of thermodynamic metastability. The hysteresis loops are experimentally

Table 10.3 *Available experimental techniques for the study of protein–water interactions*

	State of sample		*Type of information*			*Information about*	
	Solution	*Solid*	*Time-average*	*Dynamic*	*Structural*	*Water*	*Protein*
Diffraction	+	+	+	(+)	+	+	+
Spectroscopy	+		+		(+)	(+)	+
Thermodynamics	+		+			+	+
Sorption		+	+			+	+
Relaxation (NMR, esr)	+	+		+		+	+
Hydrodynamics	+			+			+
Kinetics (e.g. H^+ exchange	+	+		+			+

reproducible; they suggest conformational changes and/or slow water migration on the surface of the protein to sites of high sorption energy; *absorption* (plasticisation) must also be considered a possibility.

Despite all these complications, water sorption isotherms are often subjected to thermodynamic analyses based on the Langmuir, BET, or some other sorption theories. This makes possible the calculation of a notional monolayer coverage, a sorption energy, and other details of the sorption process. It should be remembered, however, that the assumptions on which such *equilibrium* sorption theories are based cannot possibly apply to a heterogeneous, nonequilibrium system that is subject to sorption hysteresis. Furthermore, under the experimental conditions employed, the measured relative humidity, p/p_0 is not to be equated to the water activity a_w. Indeed, a_w has no meaning where true thermodynamic sorption/desorption equilibrium does not exist, a fact that is still not widely appreciated.

Among thermodynamic properties, the heat capacity is the most sensitive indicator of overall hydration interactions, although various spectroscopic techniques do provide information about direct interactions between water molecules and specific chemical groups on the polypeptide. Figure 10.2 shows the changes in various experimental parameters as a function of the water content, expressed as g water/g protein, that occur when water vapour is admitted to a previously dried lysozyme/substrate complex. The curves labelled a, b, and c are, respectively, the infrared intensities of the carboxylate band (1580 cm^{-1}), the amide I band (1660 cm^{-1}), and the water O–D stretching frequency (2570 cm^{-1}). Curve d is the apparent specific heat, e is the diamagnetic susceptibility, f is the log peptide proton exchange rate, and g is the log enzyme activity. Curve g coincides exactly with the rotational diffusion rate of the protein, as monitored by an electron spin probe.

From the water-content dependence of the various parameters in Figure 10.2, a picture can be constructed that describes events taking place on the surface of the protein as water is admitted. This is shown in Table 10.4. Where the various properties level off near 0.4 g water/g lysozyme, water is observed to take up its 'normal' physical properties. Indications are that the first traces of water produce a 'minor' conformational relaxation of the protein resulting from the hydration and ionisation of carboxylate groups. As the degree of hydration increases, so the acidic and basic groups become saturated and water molecules diffuse from there to other sites, specifically the peptide backbone –NH and >CO groups. At this stage, a low level of enzyme activity is first observed that then increases with hydration, but only reaches its 'normal' value at 9 g water/g protein! The decrease in C at hydration levels above 0.25 is yet

Table 10.4 *Events associated with the sequential hydration (water vapour uptake) of the lysozyme/substrate complex, as monitored by the parameters shown in Figure 10.2*

Degree of hydration ($g\ H_2O/g$ *protein*)	*Site of interaction*	*Event*
0.1	Ionic side chains Glu, Asp, Lys Main chain >CO	Acids reach 'normal' pK_a values Minor conformational transition
0.2	Main chain >NH Peptide groups 'saturated'	Side chain mobility >CO, >NH stretching frequencies attain 'normal' values
0.3	Side chain –OH Water clusters, bridges	τ_{H_2O} assumes 'normal' value (~3 ps) Detectable enzyme activity

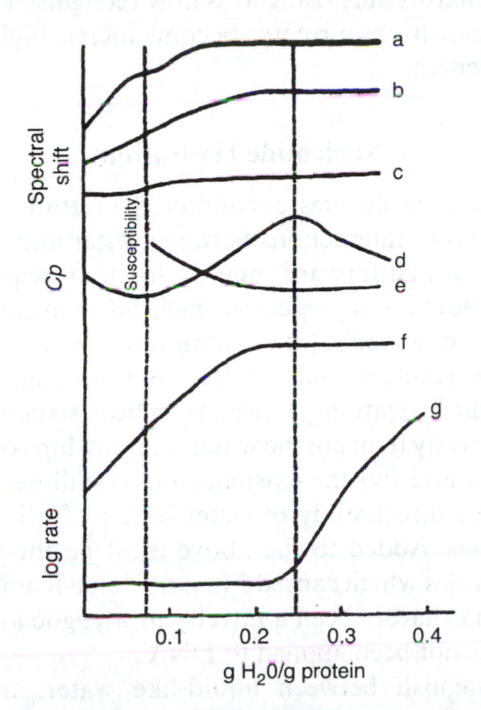

Figure 10.2 *Effect of sequential hydration on various experimental parameters of the lysozyme/substrate complex: a, b and c are respectively the infrared intensities of the carboxylate band (1580 cm^{-1}), the amide I band (1660 cm^{-1}) and the water O–D stretching frequency (2570 cm^{-1}). Curve d is the apparent specific heat, e is the diamagnetic susceptibility, f is the log rate peptide proton exchange and g is the log specific enzyme activity. Curve g coincides exactly with the rotational diffusion rate of the protein, as monitored by an electron spin probe*

(Reproduced by permission of Humana Press from F. Franks, *Protein Biotechnology*, 1993, p. 462)

to be explained; it may be related to hydrophobic hydration phenomena or water bridging of several hydration sites.

Controlled desiccation also serves to enhance the thermal stability of globular proteins; the following water contents (per g protein) will stabilise proteins against heating at 100 °C: myoglobin, 8%; haemoglobin, 11%; chymotrypsinogen, 11%; collagen, 28%; lysozyme, 15%; and β-lactoglobulin, 22%. Similar effects of desiccation are observed for the inactivation of enzymes. Thus, the heat inactivation rate constant of glucose-6-phosphate dehydrogenase (also glucose dehydrogenase and leucine dehydrogenase) in dilute solution is 0.12 min^{-1} at 50 °C but is reduced to 0.015 min^{-1} at 145 °C and r.h. = 11%.

With the more recent realisation of the important part played by glassy states in the stabilisation of proteins, to be elaborated in Chapter 13, the above results are hardly surprising. It is now recognised that, below their respective glass transitions, proteins become increasingly stable towards inactivating influences.

Nucleotide Hydration

Compared to the complexities encountered in studies of protein hydration, an analysis of interactions between water and nucleotides is, at first sight, more straightforward, mainly for four reasons: (i) the four purine and pyrimidine bases resemble each other more closely than do the 20+ protein amino acids, (ii) the long-range, electrostatic forces due to the phosphate residues and counter ions dominate the hydration structures, (iii) the hydration of regular helical structures is easier to describe quantitatively than are the water relationships of complex folded protein structures and (iv) the existence of crystalline oligonucleotides makes possible the direct study of water/base hydration geometries by diffraction methods. Added to the above must be the special scientific interest in nucleotides which caused Maxim Frank-Kaminskii to write in 1987 that 'there has hardly been a novelty or a vogue among solid-state physicists that has not been applied to DNA'.

One can distinguish between liquid-like water, trapped between packed, crystalline macromolecules, as in t-RNA, and ordered hydration shells in which water molecules occupy distinct locations that are determined by intermolecular contacts between the macromolecules. A further distinction can be made between water molecules that form nets, sheets, columns (polynucleotides) and chains and isolated water molecules that appear to act simply as 'space fillers' in the crystal (bases, nucleosides). Water also plays an important role in nucleotide–drug interactions. For instance, the deoxyguanosine–actinomycin D complex contains only 12

H_2O molecules among the 140 atoms of the complex, but 80% of the 152 hydrogen-bonds per unit cell involve water molecules. The hydrogen-bonding networks are extremely complex and resemble in a sense the characteristic networks observed in carbohydrate crystals.

The hydration of nucleic acid helices was first studied by infrared spectroscopy. DNA requires 20 water molecules per nucleotide for the maintenance of the B-form, of which more than half are spectrally modified with respect to liquid water. Six water molecules have been assigned to the phosphate oxygen atoms, while the remainder perform hydrogen-bond donor and acceptor functions with various atoms in the major and minor grooves of the DNA. In the A-form, the anionic phosphate oxygen that points into the major group is the least hydrated. The affinities for water by the various hydrophilic atoms differ considerably, ranging from a high water occupancy near the guanine O6 to a low occupancy near the guanine N3 in the Z-form.

As is also the case in protein structures, water can bridge atoms belonging to the same residue or to different residues within the same chain, but also to residues belonging to different chains, as in the triple-strand collagen fibril. Hydration interactions also fulfil the functions of relieving the strain of hydrogen-bonding between unfavourably placed atoms, on the one hand, and stabilising the helix twist on the other. This is shown in Figure 10.3 for the B-form and in Figure 10.4 for the Z-form of DNA.

Nucleotide hydration has also for long been a favourite subject for computer simulations, both of the Monte Carlo and the MD type. Average hydration structures have been calculated from the maxima in the solvent probability distribution function for water oxygens and hydrogens. Extensive water 'filaments', bridging successive phosphate groups and spanning the major groove were established. The energetics supported, not surprisingly, relative water affinities in the order phosphate > sugars > bases. As might be expected, the inclusion of Na^+ ions in the calculations led to a narrowing of the peaks corresponding to hydrogen-bonded atom pair distribution functions. The counterions were also somewhat displaced from the phosphate ions, allowing each ion to be independently hydrated (i.e. no direct ion pairing).

Developments in spectroscopic techniques have also produced an increased interest in the dynamics of DNA hydration. Caution must, however, be exercised in assigning certain molecules to primary or secondary hydration shells. In a number of cases, such divisions have produced interpretations that are hardly compatible with the laws of molecular physics and statistical mechanics. The characteristic properties of given numbers of water molecules in the primary hydration shell are

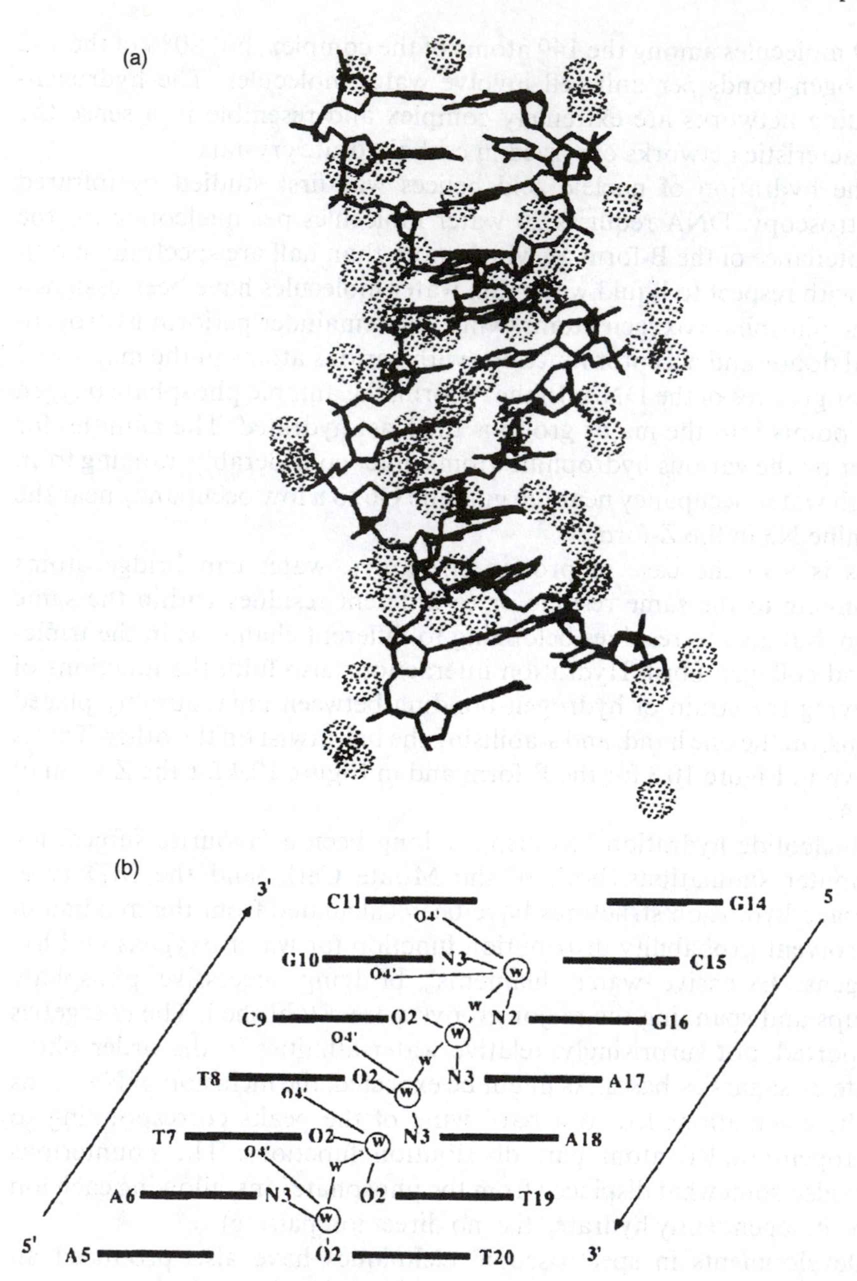

Figure 10.3 (a) *Diagrammatic representation of the spine of hydration running down the minor groove of B-DNA;* (b) *a schematic drawing, showing interactions of water molecules with oxygen and nitrogen atoms and with one another. Note the diversity of hydrogen bond orientations*
(Reproduced by permission of Cambridge University Press from E. Westhof and D.L. Beveridge, *Water Sci. Rev.*, 1990, **5**, 24)

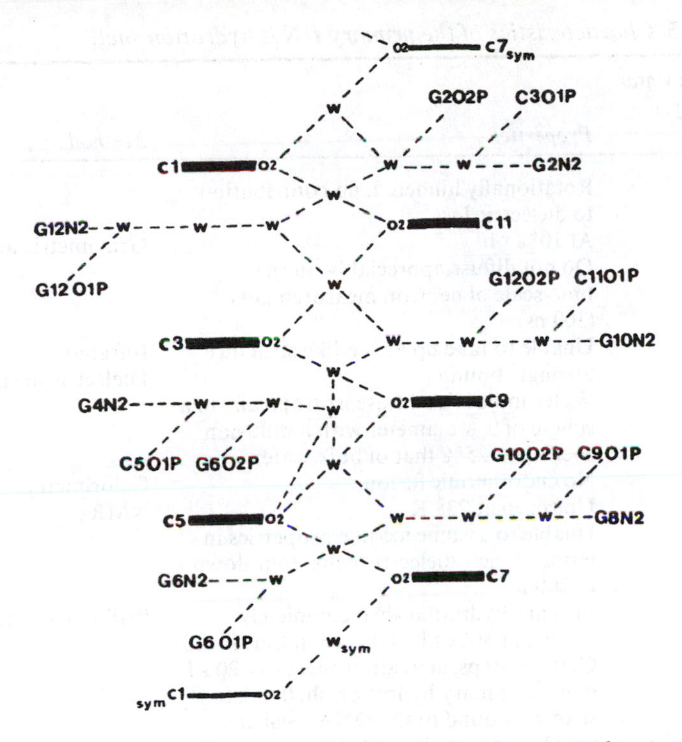

Figure 10.4 *Schematic representation of the hydration interactions in the minor groove of a typical Z-DNA hexamer*
(Reproduced by permission of Cambridge University Press from E. Westhof and D.L. Beveridge, *Water Sci. Rev.*, 1990, **5**, 24)

summarised in Table 10.5, together with the experimental method that led to the particular description. Taking the evidence as a whole, it appears that up to 10–14 waters per nucleotide have properties that differ significantly from those of liquid water and that above 20–23 waters per nucleotide, the properties of the water molecules are identical to those of pure water. In this context it is of interest to note that Na-DNA undergoes the B → A transition at r.h. > 85%, corresponding to 10–14 waters per nucleotide and is complete in the B-form at 20 waters per nucleotide. It has also been estimated that 10.5 water molecules would fit the grooves of A-DNA, whereas 19.3 water molecules would fit the grooves of B-DNA. Finally, relaxation experiments have suggested that, at 0 °C, the water in the secondary hydration shell has all the properties that are associated with undercooled water (see Chapter 12).

The first suggestions that water contributes to the stability of nucleic acid helices date from the 1950s, but it was left to the crystallographers to produce the first supporting evidence. Since nucleotide crystals contain 50–75% water at a density that approximates closely to that of liquid

Table 10.5 *Characteristics of the primary DNA hydration shell*

Number of water molecules per nucleotide	*Properties*	*Method*
1	Rotationally hindered, no contribution to dielectric loss	
2	At 10% r.h.	Gravimetric analysis
3.5	Do not diffuse appreciably on the time-scale of neutron measurements (300 ps)	
9	Unable to take up an ice-like structure	Infrared
9	Strongly bound	Dielectric measurements
9.5	Water molecules diffuse isotropically in a sphere of 9 Å diameter with a diffusion coefficient 25% that of bulk water	
10	No endothermic fusion	Calorimetry
11	Unfrozen at 238 K	NMR
14	Unable to assume ice-like properties in terms of their dielectric behaviour down to 200 K	
23	Primary hydration shell completely formed at 80% r.h., relaxation time (4–80 GHz) = 40 ps, activation energy = 20 kJ mol^{-1}, primary hydration shell most strongly bound to the DNA itself and poorly connected to external waters	Brillouin scattering
35	Interact with DNA (weakly or strongly)	Dielectric measurements

water, a large majority of water molecules participate in hydration complexes with anionic oxygens by direct hydrogen-bonding, with mobile ions by ion–dipole interactions or form quasiclathrate cages around apolar groups. This diversity of hydration behaviour then gives rise to the complex hydration network. The different environments of water molecules in sugar–water–base and base–water–base bridges and those in the proximity of ions also give rise to the observed complex dynamic behaviour.

There is still a tendency to think of crystallographically ordered water as 'the' hydration, neglecting the facts that *all* water molecules are subject to more or less rapid exchange with water in the bulk and that not all water molecules exist in the ordered state. As regards the role of water in promoting or maintaining a particular biopolymer configurational state, it cannot be emphasised too strongly that stability is a matter of the free energy *of the whole system*, and not of the apparent structure. Structural biology literature, especially as it pertains to nucleic acids, suffers from some confusion between structure and thermodynamics, especially where labile, non-covalently bonded hydration structures are involved. Water forms an integral part of all ordered biological structures, helical and otherwise, a factor that must be taken into account in any adequate description of such a system. This *caveat* is of particular importance for descriptions of nucleotides, because of their highly charged nature.

Lipid Thermotropism and Lyotropism

Lipids are included here because, together with the biopolymers, they constitute the basic building blocks of biological structures. They differ from the other, polymeric species described in this chapter because, by comparison, they are 'small' molecules. The structures formed by lipids are stabilised by noncovalent interactions, resulting from hydration and steric factors. In the main they are derivatives of glycerol in which one, two or all three –OH groups have been replaced by acyl chains and ionogenic or polar groups. Some other molecules that are also often classified as lipids include steroids, e.g. cholesterol.

Within the context of this chapter, pride of place goes to those polar lipids that are the main constituents of biological membranes. They display a complex phase behaviour that is sensitively dependent on temperature (thermotropism) and hydration characteristics (lyotropism). Examples of the calorimetric detection of typical thermotropic phase transitions are shown diagrammatically in Figure 10.5(a) for the zwitterionic dipalmitoylphosphatidylcholine (DPPC). At low temperatures the lipid exists in a quasisolid structure (gel phase) with tilted, but orientationally ordered acyl chains. An increase in temperature introduces a degree of disorder and finally produces a well-defined transition to a liquid crystalline phase in which the acyl chains acquire rotational freedom, while the characteristic bilayer structure is, however, maintained. In contrast, the polar lipid dioleylphosphatidylethanolamine (DOPE), shown in Figure 10.5(b), exists in a hexagonal gel structure with the acyl chains perpendicular to the bilayer structure. The same transition to a liquid crystalline phase, as observed for DPPC, occurs, although at a much lower temperature, reflecting the unsaturation in the oleyl chain. A second major phase transition takes place and gives rise to an inverted hexagonal phase in which the lipid forms long, water-filled cylinders, with the polar headgroups facing inwards (towards the water).

The pronounced effect of water content on the phase transitions (lyotropism) of DPPC and didodecylphosphatidylethanolamine (DPPE) is illustrated in Figure 10.6. It is found that no freezing transition of water can be detected until the mole ratio water:DPPC exceeds 20:1. Such unfrozen water is believed to be associated with the headgroup. X-ray and NMR studies of lipid/water mixtures have confirmed the location

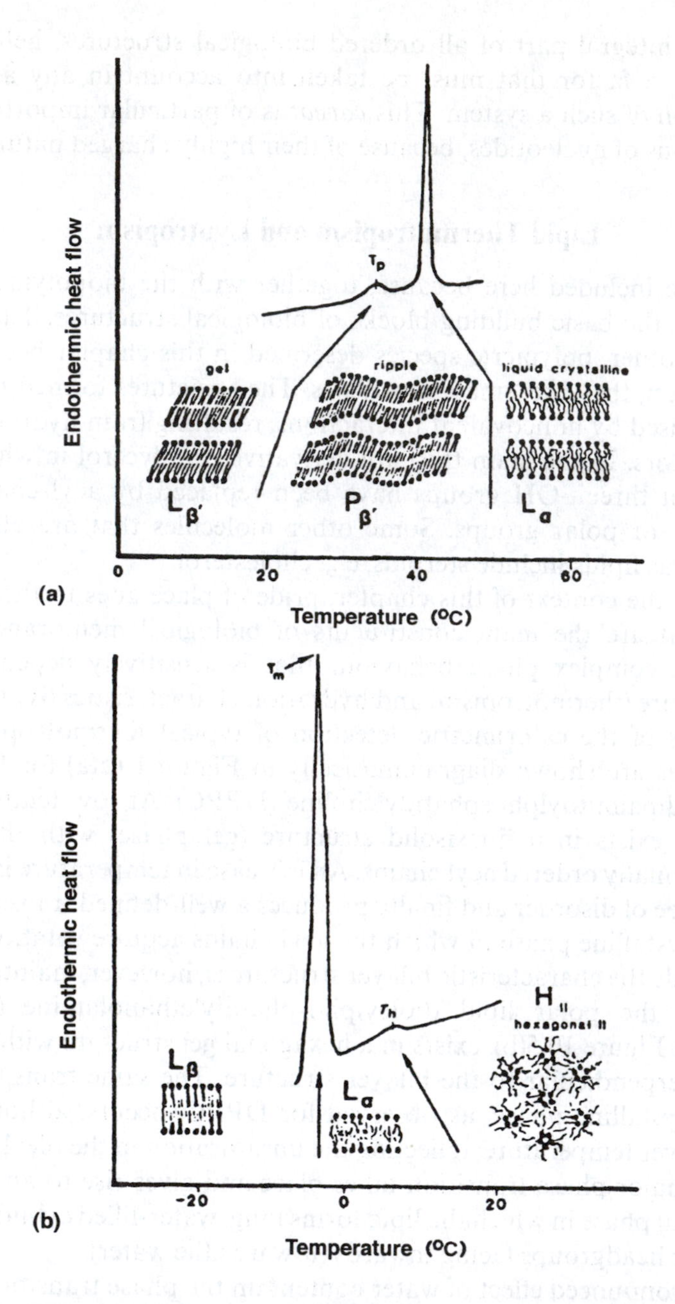

Figure 10.5 *DSC scans of* (a) *hydrated DPPC vesicles and* (b) *hydrated DOPE, showing also the bilayer structures of each phase. The phase transition temperatures are indicated by endotherms; for details, see text*
(Reproduced by permission of Cambridge University Press from J.H. Crowe and L.M. Crowe, *Water Sci. Rev.*, 1990, 5, 1)

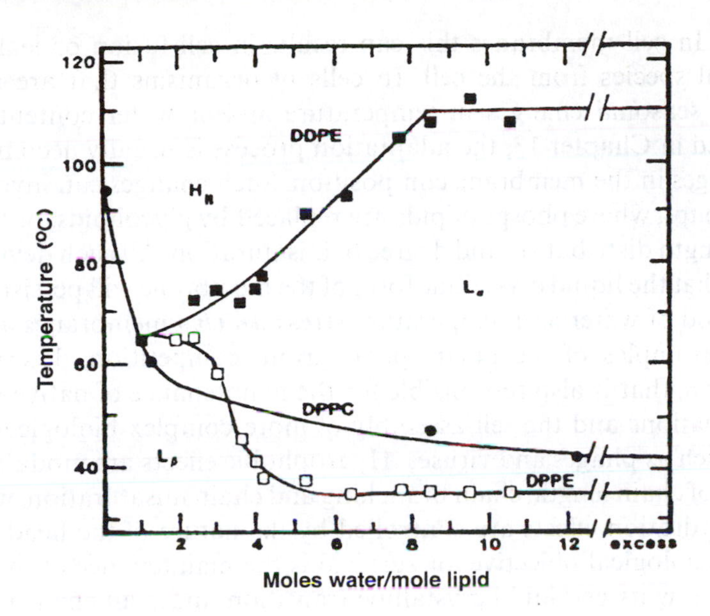

Figure 10.6 *Phase diagrams of DPPC, DOPE and DDPE as function of water:lipid mole ratio*
(Reproduced by permission of Cambridge University Press from J.H. Crowe and L.M. Crowe, *Water Sci. Rev.*, 1990, **5**, 1)

and restricted mobility of water molecules. Thus, 2 mol of water per mole DPPC are hydrogen-bonded to the phosphate groups in the same half bilayer and form phosphate–water ribbons that link adjacent phosphatidylcholine groups, but they are also hydrogen-bonded across the double layer to other water molecules.

The gradual hydration of ordered lipid phases from an almost dry state therefore proceeds in a sequential manner, as was already demonstrated for the hydration of 'dry' proteins in Figure 10.2. The interaction of the lipid headgroups with water tends to increase the distance between lipid molecules in the bilayer. This has an indirect effect on the bilayer stability, because it also serves to separate the acyl chains and thus to weaken their interactions with one another, giving rise to chain flexibility. In vesicles or membranes, increasing hydration levels will introduce a degree of elasticity and deformability.

Natural membrane systems are composed of lipid mixtures, with a distributions of head groups, acyl chain lengths and degrees of saturation. Changes in temperature, especially where accompanied by desiccation, have pronounced effects on the integrity of bilayer lipid phases, with major physiological consequences. Thus, dehydration may be accompanied not only by thermotropic and/or lyotropic phase transitions but also by two-dimensional phase separations within the plane of the bilayer. In cell membranes this can results in cell fusion or leakage of chemical species from the cell. In cells of organisms that are able to tolerate seasonal changes in temperature and/or water content, to be discussed in Chapter 13, the adaptation process is usually accompanied by changes in the membrane composition. Such changes can involve the headgroups, where phospholipids are replaced by glycolipids, or the acyl chain length distribution and degree of unsaturation. All such devices will ensure that the liquid crystalline form of the membrane will persist during the period of water and temperature stress. *In vivo* membranes are thus classic examples of the polar/apolar group competition, described in Chapter 8, that is also responsible for the maintenance of native protein conformations and the self-assembly of more complex biological structures, such as phages and viruses. Hydrophobic effects are modulated by control of chain length, chain branching and chain unsaturation, whereas direct hydration effects are controlled by the nature of the headgroups. The physiological objective for survival is the maintenance of the membrane below its gel/liquid crystalline transition under all environmental conditions.

Oligo- and Polysaccharide Hydration

Sugars and their derivatives play a central role in metabolic reaction cycles, while their oligomers and polymers fulfil biological functions that are almost as diverse as those of proteins. Carbohydrates are also used extensively in many branches of chemical technology, some of which are shown in Table 8.1. In fact, 95% of all water-soluble polymers used by industry are carbohydrate-based, mostly chemically modified in various ways.

The chemistry, especially the stereochemistry, of carbohydrates is vastly more complex than that of peptides or nucleotides, and their affinity for, and similarities to, one another and water make experimental studies of their hydration behaviour difficult. In what was probably the first critical review of the subject, Suggett wrote in 1975 that water '. . . was treated implicitly as the universal, inert filler'. A more recent review by the present author concluded that the situation had hardly changed by 1990, and this is still true today. The complex solution chemistry of sugars has been touched upon in Chapter 8. It arises partly from the labile nature of sugars, which are able to exist in different stereoisomeric forms that are freely interconvertible in solution, and partly from the physico-chemical similarity of their –OH groups and those of water.

Some information on interactions between PHCs and water might be gleaned from a study of crystal structures, especially crystalline hydrates.

Because of the affinity of PHCs for water, it might be expected that a variety of hydrates might exist, but that is not the case. Few PHC hydrates have been reported; they range from monohydrates (glucose, lactose) to pentahydrates (raffinose). The possibility exists, however, that some hitherto undiscovered hydrates might exist, but only over a limited temperature range. Thus, a previously unknown hydrate of mannitol of limited temperature stability was recently identified during the freeze-drying of a mannitol solution.

Where crystal structures are available, they are found to be extremely complex. This is well illustrated by a study of the raffinose–water system. In the crystal, the trisaccharide molecules [galactopyranosyl-glucopyranosyl-fructofuranoside] are arranged in the form of a complex system of extensively hydrogen-bonded chains. The structure analysis of the pentahydrate demonstrates that *all* oxygen atoms in raffinose *and* water molecules participate in hydrogen-bonding. The sugar chains intersect only at the locations of water molecules, with all water molecules involved in at least three hydrogen bonds. Four of the five water molecules form two acceptor and two donor hydrogen bonds, with only one molecule acting as double donor and single acceptor. A simplified water hydrogen-bonding scheme at the chain intersection points is shown in Figure 10.7, illustrating that water molecules at chain intersections form hydrogen bonds with sugar residues but also with other water molecules. The actual sugar –OH sites to which specific water molecules are bonded are not shown.

By analogy with many inorganic systems that exhibit multiple stable hydrates it might be inferred that progressive drying of the pentahydrate should give rise to a succession of lower hydrates, eventually leading to the anhydrous crystalline sugar. This is found for monohydrate sugars, but the situation for sugars which form higher hydrates appears to be more complex. If the rate of desiccation exceeds the rate required for the necessary hydrogen-bonding rearrangements to occur in the increasingly

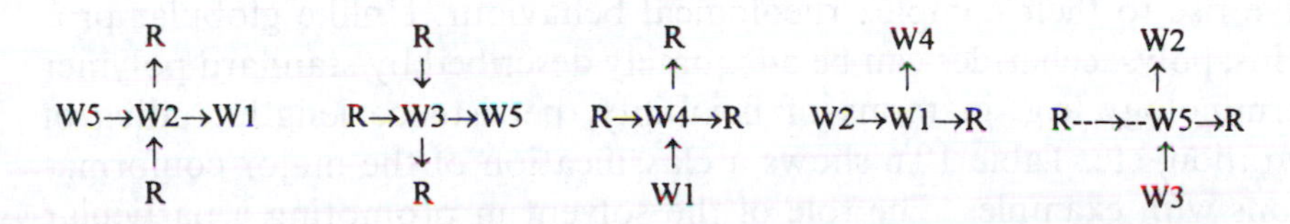

Figure 10.7 *Water-centered intermolecular hydrogen-bonding patterns in crystalline raffinose pentahydrate. Arrows indicate the direction of proton transfer from donor to acceptor. A broken line indicates a long (weak) hydrogen bond*

viscous, supersaturated sugar–water mixture, then it is probable that bond mismatches will occur and become frozen into the network, giving rise to amorphisation. An analogy can be drawn with the well-known residual entropy of ice which results from the 'freezing-in' of hydrogen atoms in random positions along the O–O axes.

Conceptually, the removal of water molecules from a hydrate crystal might proceed in two distinct ways, probably depending on the differential hydrogen-bonding energies. Either groups of specific water molecules (e.g. all those designated as W1 in Figure 10.7) are removed sequentially, or water molecules are removed from the network randomly. The progressive removal of water molecules from the lattice must in any case lead to a perturbation of the hydrogen-bond topology. This might be accompanied by one of several possible 'repair' mechanisms: either new sugar–sugar bonds will be formed to replace the sugar–water bonds, giving rise to a new crystal structure, or a new network will form, but in a random manner, thus producing amorphous domains within the remaining pentahydrate crystal structure. The latter mechanism is more likely to occur if water had previously been removed from the crystal in a random manner. Evidence exists that desiccation leads to the sequential loss of specific water molecules from the lattice, at least down to a notional trihydrate stoichiometry. Further desiccation, as performed experimentally, may then become too 'fast' for new, ordered lattices to become established, with the practical result of progressive amorphisation, a phenomenon that is discussed in more detail in Chapter 13.

The above discussion demonstrates that our understanding of PHC–water interactions, even in the solid state, is still rudimentary, apart from the availability of spatial atomic coordinate data. On the other hand, that oligo- and polysaccharides are sensitively attuned to their solvent medium becomes apparent from a study of their molecular shapes in solution. Although their conformations may not display the same degree of specificity as those of native, folded proteins, they nevertheless can be made to undergo structural transitions between three extreme forms, compact spheres, random coils and rigid rods, by changes in the solution environment, e.g. pH, temperature, cosolvents. Such changes give rise to their complex rheological behaviour. Unlike globular proteins, polysaccharides can be adequately described by standard polymer terminology, e.g. in terms of flexibility, persistence length, radius of gyration, etc. Table 10.6 shows a classification of the major conformations with examples. The role of the solvent in promoting a particular macromolecular shape remains to be explained.

Glycobiology forms a particularly important subdiscipline of carbohydrate chemistry. It is concerned with the synthesis and functions of

Table 10.6 *Conformational diversity of polysaccharides in solution*

Nature	*Examples*
Extra-rigid rod	Schizophyllan, xanthan
Rigid rod	Alginate, chitosan (highly deacylated), pectin (highly de-esterified)
Semi-flexible or asymmetric coil	Xylans, cellulose nitrate, methyl cellulose, mucin glycoproteins
Random coil	Pillulan, dextran, guar, locust-bean gum, konjac mannan
Globular or highly branched	Amylopectin

oligosaccharide residues attached to proteins. The sugar groups are believed to be responsible for several physiological functions, among them cell–cell recognition. It is held by some, that *all* proteins originate as glycoproteins at the time of synthesis and that the oligosaccharide residues are in some cases removed *in vivo*, and in other cases during 'clumsy' downstream processing. The *in vivo* glycosylation process is complex, and the resulting oligosaccharide residues take on a wide range of structures. In line with what we know about sugars in solution (Chapter 8), it can be safely assumed that the glyco residues also interact with water. Their particular 'native' conformations presumably give rise to their ability to recognise each other, and these conformations are to some extent due to the hydration interactions. We are, however, not yet advanced enough in our understanding to take such second-order (?) effects into account in present-day experimental and theoretical studies of glycoprotein structures and functions. The state of our knowledge is approximately what it was for globular proteins some fifty years ago, when all calculations pertaining to folding and unfolding were performed on peptide chains in the absence of the solvent medium, i.e. *in vacuo*, with hindsight surely a somewhat unrealistic and unprofitable pastime. The experimental and computational problems associated with the inclusion of hydration effects into studies of carbohydrate conformational behaviour are of a different order of magnitude. Even more complex will be the calculation of hydration interactions in 'mixed' biopolymers, such as immunoglobulins, glycoproteins, glycolipids, nucleoproteins, etc. It is nevertheless to be hoped that attempts to include the solvent contribution in such studies will eventually be reported, because, without due consideration of such effects, any interpretations of interactions, configurations and functions must lack some credibility.

Chapter 11

Water in the Chemistry and Physics of Life

The diverse roles played by water in controlling life processes are subjects that are rarely to be found in biology teaching. Most courses, whether at school or university, do not even mention water. Its involvement can be studied at various levels. Thus, water participates in most biochemical reactions. It also functions as lubricant and cleaning fluid, transporting both nutrients and waste products to and from target organs. Most important, for much of the living history of our planet (ca. 3.8×10^9 years) water was the only natural habitat for life. Indeed, life on so-called dry land developed comparatively recently (4×10^8 years ago). Even now, the advanced life forms, mammals, still begin their development in the aqueous environment of the womb. We shall presently return to a brief comparison of life in water and life on dry land. Suffice to mention that the emergence of species onto dry land has had to be paid for in terms of major structural (skeletal) changes and energy-consuming processes to maintain a correct water balance.

The Physiological Water Cycle

Within the overall hydrological cycle, described in Chapter 1, can be found several subcycles. Of these, the physiological cycle illustrates water turnover by living organisms. Thus, plants require water for photosynthesis and lose water by transpiration and evaporation, while animals ensure their water supply mainly from food and drink, which is balanced by a water loss due to perspiration, evaporation and excretion. Water housekeeping is thus an important facet in ensuring health and the optimum functioning of all living species. Many pathological conditions can arise from disturbances in the correct water management by the organism.

Water Biochemistry

At the outset of any discussion of the biochemistry of water, it is emphasised that, unlike most chemical-engineering processes, which aim at maximisation of yield at a minimum cost, processes occurring in nature are characterised not by a high efficiency but by the economy with which chemical energy is utilised. It follows that in living organisms, which might consist of up to 98% of water (jellyfish), this liquid fulfils a function over and above that of an inert substrate. It is much harder to elucidate the exact role(s) of water in the complex chemistry of life processes. Apart from acting as a proton-exchange medium, water moves through the organism and functions as lubricant in the form of surface film and viscous juices, e.g. dilute secretions of mucopolysaccharides.

Water participates in four major types of biochemical reactions: oxidation, reduction, condensation and hydrolysis. There are many other biochemical reactions in which water splitting or synthesis form important stages but where the exact mechanisms are still mysteries. Essentially, oxidation is the removal of electrons from O^{2-} to form O^- and eventually gaseous oxygen. Similarly, protons are reduced to gaseous hydrogen by the addition of electrons. The oxidation of water to oxygen constitutes one of the important component reactions of photosynthesis. The whole process occurs in three stages: a photochemical excitation of the photosynthetic pigments (chlorophylls) causing a release of electrons, the electron-transfer reactions, leading to the reduction of nicotinamide adenine dinucleotide phosphate (NADP) to $NADPH_2$, and the 'biochemistry', involving the conversion of CO_2 to carbohydrate. The electron-transfer stage is the least well understood; it includes the oxidation of water and the synthesis of adenosine triphosphate (ATP). Schematically the oxidation of water can be written as

$$2H_2O = O_2 + 4H^+ + 4e^-$$

but more correctly the reaction is

$$4H_2O = O_2 + 4H^+ + 2H_2O + 4e^-$$

However, little is known about the detailed mechanism whereby water is oxidised to gaseous oxygen, a reaction that is not easily accomplished by conventional chemical means.

Other reactions involving the splitting of water include the conversion of fumaric acid (*trans*-butenedioic acid) to malic acid (2-hydroxybutanedioic acid), one of the steps in the citric acid cycle, a reaction sequence by means of which three-carbon acids are oxidised to CO_2 and water. According to standard textbooks, water (i.e. H and OH) is added directly across the double bond in a single step in the presence of the enzyme fumarase. This is surely a remarkable statement, but it illustrates the cavalier manner with which water is treated by chemists, and, more particularly, by biochemists. Another example will serve to bring this point home. The citric acid cycle (also known as the tricarboxylic acid cycle) consists of eight consecutive reactions, each controlled by a specific enzyme. In several places water is split in a manner similar to that shown above for photosynthesis, and the electrons are utilised to reduce NAD. Some popular and famous biochemistry textbooks present reaction sequences and diagrams in which the hydrogen budgets do not balance, i.e. the number of protons entering the cycle through the splitting of water does not equal the number of protons leaving the cycle. In other words, the decomposition and synthesis of water are not treated stoichiometrically, and yet these reactions fulfil a central role in the synthesis of ATP. Without proper 'water book-keeping' a perceptive student might find it difficult to understand how one molecule of glucose (12H) could possibly supply 12 pairs of protons for the production of 36 molecules of ATP in the citric acid cycle.

The citric acid cycle is cited as an example of negligence in the stoichiometric representation of metabolic reaction sequences, but many examples abound in the biochemical literature, including, regrettably, undergraduate texts. It is clearly a case of 'familiarity breeds contempt'. To biochemists the chemical transformation of organic molecules in metabolism and synthesis takes precedence over second-order effects (?) and 'proton book-keeping', as related to the oxidation/reduction of the common solvent medium. It is unlikely, however, that *correct* mechanisms for complex metabolic reaction sequences can be established without taking such effects into account. It is hardly an exaggeration to claim that biochemistry is primarily the chemistry of water.

Examine, for example, the production of water through the combustion of carbohydrates, as it takes place in the mitochondrion. The average human adult has a daily water turnover of approximately 4% of the body weight: 2.5 kg, of which 300 g is produced endogenously by the oxidation of carbohydrate. The remainder is absorbed by the intake of food and drink, while the loss is accounted for by perspiration, transpiration and excretion. Taking glucose as the primary nutrient, then the following equation would summarise its direct oxidation:

$$C_6H_{12}O_6 + 6O_2 = 6CO_2 + 6H_2O$$

The synthesis of 300 g of water by this mechanism is accompanied by the liberation of energy to the amount of 7600 kJ, enough to raise the body temperature by 26 °C. Actually this enthalpy is converted into chemical energy which is stored in the form of ATP. A more correct form of the above equation is then

$$C_6H_{12}O_6 + 6O_2 + 38ADP + 38P_i = 6CO_2 + 6H_2O + 38ATP$$

where P_i represents inorganic phosphate and ADP is adenosine diphosphate. The 300 g of water produced by this reaction are therefore accompanied by the synthesis of about 100 mol ATP, which is stored and provides the energy requirements of the many physiological functions of the body.

Unlike a chemical reactor, the living cell operates under isothermal conditions, with a maximum conservation of energy. The stored energy produced by the oxidation of glucose is utilised in several ways:

for the synthesis of other molecules,
as osmotic energy, measured as a pressure difference,
as energy of motion, i.e. muscle contraction,
as electrical energy, i.e. nerve stimulation, and
as radiation, e.g. bioluminescence.

Even written in the above form, the glucose oxidation equation is a gross oversimplification of the real reaction sequence, because the actual metabolic fuel is not glucose but NADH which is oxidised to NAD, with each mole NADH producing 1 mole water. The total *in vivo* oxidation of glucose and the simultaneous synthesis of ATP (and water) takes place in six stages, with a stepwise removal of two electrons, each of the six steps giving rise to a carboxylic acid, which is then removed as CO_2. The two electrons are stored in an NAD molecule, which is itself, reduced to NADH:

$$NADH \rightarrow NAD + 2e^-$$

$$2e^- + \tfrac{1}{2}O_2 + 2H^+ \rightarrow H_2O \qquad \Delta H = -217\ \text{kJ mol}^{-1}$$

The oxidation of glucose and the simultaneous synthesis of ATP (and water) take place in a cascade of 14 steps, each controlled by an enzyme. Water participates in each step. The *in vivo* mechanisms and rates of all such coupled reactions have, in the course of evolution, become sensitively attuned to the properties of water, such as its ionisation equilibrium and its hydrogen-bonding pattern. Even small changes in any of these properties, e.g. the substitution of ^1H by ^2H, causes chaos to the coupling between biochemical reactions, and hence to the viability of the organism. Heavy water is thus toxic to all higher forms of life.

The production of 300 g of water also requires 185 litres of oxygen (approximately 40% of the total daily oxygen requirement), which the lung extracts from air with an efficiency of 14%. Since air contains 21% of oxygen, the lung must process some 6300 l of air daily, in order to generate the necessary supply of oxygen. By extending such calculations, it can be shown that, just to supply the cells with enough oxygen for the daily combustion of glucose, the heart must pump 7000 l of blood around the vascular system. The quantitative aspects associated with a daily mitochondrial 'factory' production of 300 g water are illustrated in Figure 11.1.

Other organs that control physiological processes associated with water housekeeping include the kidney. The 'normal' daily 1.4 l of excreted urine is produced by the concentration of 180 l of dilute urine. The remaining water is returned, purified, to the body. This degree of concentration to produce pure from contaminated water requires a large amount of energy; qualitatively it resembles the industrial process of water desalination. The excreted urine contains ca. 50–75 g of solids, 20–30 g of which is urea. Another function of the kidney is to regulate the quantity of body water and its salt content. Aqueous juices, totalling 8 l, are also produced daily and controlled by the salivary glands, pan-

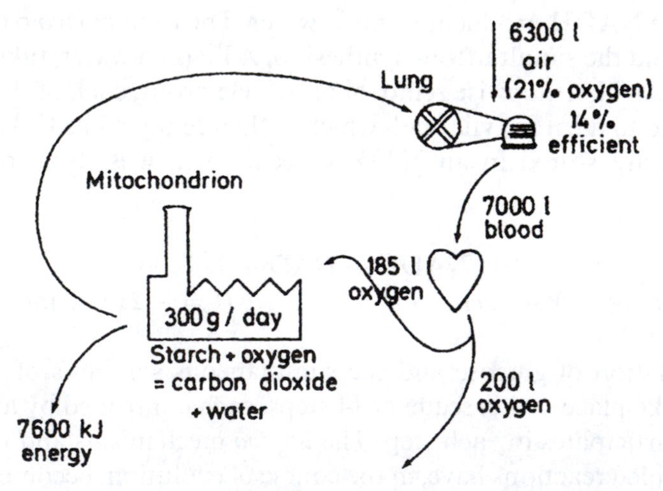

Figure 11.1 *The daily human 'factory' output and consumption, associated with the mitochondrial generation of 300 g of water*

creas, intestines, gall bladder and liver. Each heartbeat pumps 70 ml of blood, so that with 70 beats per minute, the heart supplies the 7000 litres required for the generation of energy by the mitochondria, as discussed above.

Before leaving this brief account of biochemistry, as seen from the point of view of water, it must be stressed that all other biosynthetic processes, e.g. protein and nucleotide synthesis and metabolism, are similarly coupled to the chemistry of water. If, however, the combustion of proteins, lipids and carbohydrates to yield ATP could proceed in an unbalanced manner, then animal life could only continue for a few days. The replacement of the basic nutrients is performed by plants which, by means of photosynthesis, generate the energy in the form of those chemicals that are utilised by the animals, mainly carbohydrates, but also proteins and lipids, to a limited extent. The key reaction here is the *oxidation* of water, i.e. the converse of the reaction used by the glucose consumers.

All plants on earth turn ca. 5×10^{10} tons of carbon into carbohydrate annually. This requires the splitting of ca. 8.5×10^{10} tons of water, of which marine plants contribute 50%. The annual energy requirement is 4×10^{15} kJ. Actually the plants absorb solar energy of the order of 10^{18} kJ, because of the low efficiency of the photosynthetic process and the immense energy consumption required for the evaporation of water. Even so, such quantities amount to only 0.1% of the solar energy radiated to earth.

Of all the raw materials required for the cyclic processes of life, water is the only one in abundant supply. Both nitrogen and carbon dioxide supplies are much more finely balanced in the ecosphere and can easily be upset by human interference, sometimes with disastrous consequences.

Physics of the Natural Aqueous Environment

The physical properties of water must have played an important role in evolution. This can be most convincingly illustrated by comparisons of water and air as the natural environments of all living beings. Properties which have had the most profound influence on the development of life include density (weight, pressure, fluid dynamics), viscosity, diffusion, thermal and electrical properties, surface tension and propagation of light and sound. Outstanding puzzles of water management in the plant kingdom include: how does water reach the top of a tree and what are the forces that produce the motion of water within plant cells and in the xylem and phloem of intact plants?

It is beyond the scope of this book to treat all the above in detail, but as an illustration let us consider the impact on life of one very basic physical property of water: density, and its change with temperature. The two density abnormalities of fresh water are its non-monotonic temperature dependence and its decrease upon freezing. Everybody is familiar with the consequences of the density maximum at 4 °C for aquatic life. What is not so well known is that seawater with a typical salinity of 35 g kg^{-1} differs from fresh water in that the temperature of maximum density lies below the freezing point. Thus, in the ocean, the coldest water *is* the densest water. The density of ocean water is affected by depth (pressure) and salinity in a complex manner, which impacts profoundly on the buoyancy of aquatic organisms. From a comparison of the body densities of terrestrial and aquatic organisms it can be concluded that the terrestrial organisms in the lineage leading to present-day mammals did not acquire a mechanism for substantially altering their density when they left the oceans for dry land.

Many plants and animals have the form of upright columns and therefore have a tendency to fall over. The critical height of a stable column is

$$h_{max} = 1.26\frac{Er^2}{g\rho_e} \qquad (11.1)$$

where E is the stiffness (Young's modulus), r is the radius and ρ_e is the effective density, that is the density difference between the material and the surrounding medium. A tree of 1 m radius could grow to a height of almost 130 m, which is higher than the tallest known tree. Applying the equation to a water environment gives the curious result of a negative maximum height. Because wood is less dense than water, the buoyant force on it more than offsets its weight and it floats. Buoyant water plants do not have stability problems, but some aquatic plants, e.g. marine macroalgae, have densities of approx. 1080 kg m^{-3}, corresponding to $\rho_e = 55$ kg m^{-3} in seawater. They also have a low modulus. The above equation predicts that for a radius of 1 cm, the plant would become unstable when its height exceeds 1.5 m. In air, such a plant could only grow to a height of 0.5 cm. Intertidal algae therefore stand upright when the tide is in and lie on the shore when the tide is out. The impression has probably now been created that water is a more favourable environment for the maintenance of stability than air, but such an advantage is more than offset by fluid-dynamic forces, which affect columns in water much more profoundly than in air.

The effects of density are also found in swim-bladders, membrane-enclosed gas pockets within the body cavity of some fish species whose densities exceed that of seawater. Although the pumping of gas from blood, where it has a low pressure, to the bladder, where it has a high pressure, requires time and energy, the mechanism is effective in preventing the fish from sinking.

Hydrodynamic forces that are directly proportional to density include pressure drag, lift and acceleration. Locomotion (thrust) is produced by lift and drag. Some animals, e.g. dolphins, can break the normal link between thrust and drag by launching themselves into the air where they experience a reduced drag. The 'cost' of locomotion can be expressed by SI units, J (kg m)$^{-1}$, but more usefully by the volume of oxygen required to propel 1 kg of body mass by 1 m. The consumption of 1 l of oxygen corresponds approximately to 21 kJ, depending on the material (carbohydrate or lipid) that is being oxidised. It is hardly surprising that it costs three times more, in terms of oxygen consumption, to transport a body by walking or crawling than it does by swimming, mainly because a buoyant fish does not need to expend energy to maintain its upright position against gravity.

As a final example of the biological consequences of density let us consider blood pressure and turgor pressure. The blood vessel in a fish can be approximated to a liquid-filled tube immersed in water. Any change in the hydrostatic pressure is counteracted by a change in pressure in the surrounding water. The arterial blood pressure is almost the same everywhere in the body, and the arteries have to resist only the pressure created by the heart. In terrestrial animals the situation is quite different. The external air pressure exerted on the vascular system does not vary nearly as much with the vertical position as does the internal blood pressure. The arteries and veins in the feet are therefore hypertensive and must be able to cope with this higher pressure.

Water in the xylem of vascular plants forms a continuous column from the roots to the leaves. The roots therefore experience a much higher pressure than the crown. The water is supported by root (osmotic) pressure and by the fact that the top of the water column does not allow air to enter. The pressure starts at a high value in the roots, reaches the ambient value a few metres above ground and decreases further with increasing height. Much of the water is then at a *negative* pressure, and xylem has the opposite problem of an artery because its walls must resist a strong inward force.

The whole subject of water flux in plants or blood in vascular systems must be considered in terms of how pressure gradients (Δp) in a tubule of length l affect fluid flow. The average volume flux J is given by the well-known Poisseuille equation

$$J = \frac{\pi r^4 \Delta p}{8\eta l} \tag{11.2}$$

where η is the dynamic viscosity.

The power P required to pump fluid through the tube is given by $J(\Delta p)$:

$$P = \frac{\pi r^4 \; (\Delta p)^2}{8\eta l} \tag{11.3}$$

A typical xylem cell has a radius of 20 μm. When a tree is actively respiring, the water velocity is ca. 1 mm s^{-1}. This flow would require a pressure gradient of ca. 2×10^4 Pa m^{-1}. For a tree 60 m high this translates into a pressure difference of 20 atm across the length of its vascular system (in addition to a hydrostatic pressure gradient of 10^4 Pa m^{-1}) and a required power of 200 W per m^2 of vessel tubes. This energy is provided by the mechanism of evaporation from the leaves.

One further factor needs to be considered in a discussion of water transport in plants: surface tension. The material of the xylem tubes is effectively wetted by water (zero contact angle), so that capillary forces would maintain a column of water of height h, as given by

$$h = \frac{2\sigma}{r\rho_e g} \tag{11.4}$$

where σ is the surface tension (0.073 J^{-2}), r is the radius of the xylem tube (typically 20 μm), and g is the gravitational constant (9.8 m s^{-2}). The capillary rise h is then of the order of 74 cm, so that plants below that height could deliver water to their leaves by capillary action alone.

A more sophisticated mechanism is required for maintaining the water flux in tall trees. The xylem tube is subdivided at the top end by leaf cells that are permeable to water, but each with its own water–air interface. Their radii are typically 5 nm. Substituting this value in equation 11.4, gives $h \approx 3.4$ km. The negative pressure at the top of such a column would be -300 MPa, much greater than the cohesive strength of water. The tensile strength of water would limit the height of a tree to ca. 280 m, still more than double the height of the tallest trees found. The factors which limit growth are not yet well understood.

The above discussion of density effects might have given the impression that marine life has fewer problems to cope with than have terrestrial species. There are, however, other factors that make life in the ocean uncomfortable. The limited supply and slow diffusion of oxygen make breathing in water hard work. Inhabitants of the ocean also exist in a state of osmotic disequilibrium, so that energy is required to maintain the body fluids at the correct osmotic concentrations. Most bony fish achieve this by drinking seawater and excreting excess salt. They can only produce urine that is in osmotic equilibrium with their body fluids, so that the excess salt is secreted by special cells located in the gills.

Actually, mechanisms to cope with high salt concentrations are not confined to animals. Whereas most plants are sensitive to soil salinity, and specific Na^+ toxicity is a well-established phenomenon, there are some plants which assimilate salts. The relationship between growth, ion uptake and accumulation, and the importance of such processes to the mechanism of salt tolerance, are still somewhat unclear. The increasing losses of arable land due to secondary salination (i.e. caused by irrigation with low quality water) should provide some urgency to studies of salt tolerance in crops.

Chapter 12

'Unstable' Water

Liquid Water outside its 'Normal' Temperature Range

In the preceding chapters discussions have centred on water in its stable states, i.e. stable crystalline polymorphs of ice and the stable liquid, as it exists at 'normal' temperatures and pressures. It has, however, long been known that water can exist outside these limits; undercooled water was already being studied during the eighteenth century, while studies of superheated liquid water received an impetus over the past 30 years. The important ecological and technological implications of vitrified (glassy) water were first realised some twenty years ago, and since then the vitrification of liquid water, and particularly of aqueous solutions, have become subjects of major scientific, ecological and technological interest. The temperature range over which liquid water can exist certainly extends from −41 to ca. 280 °C. If vitrified liquid water is included, then the theoretical low temperature limit must be extended to −273.16 °C. (0 K). It must be understood that outside the normal temperature range at atmospheric pressure (0–100 °C), the liquid is thermodynamically unstable, so that sudden transitions to the stable state (ice or steam) are governed solely by the laws of kinetics, i.e. probability.

Undercooled Liquid Water

The concept of *low* temperature is relative. To a metallurgist, a low temperature might be one at which liquid metals are quenched and tempered, whereas to many a physicist low temperature is associated with quantum effects, such as superconductivity. The two temperature ranges differ by some 800 degrees.

It must surely be a coincidence that the freezing point of the liquid that is essential for life lies almost exactly at the centre of the temperature range normally associated with life on this planet. Subzero temperatures are very common in the terrestrial environment, and in order to survive many living organisms have developed the means for coping with the twin effects of low temperature and freezing.

Although the equilibrium freezing point is exactly defined as the temperature at which the free energy curves of the solid and liquid states intersect, the phenomenon of undercooling is common and well documented. There is a practical limit to the degree of undercooling that can be achieved, at least at moderate cooling rates, and the lower the temperature of the undercooled liquid the more susceptible it becomes to chance freezing. The relationship between the lowest temperature of undercooling (T_h) and T_f, the equilibrium freezing point, appears to be remarkably constant for molecular liquids: $T_h/T_f \approx 0.80 \pm 0.05$.

The physical properties of liquid water change dramatically and discontinuously at its freezing point. This is illustrated by the temperature dependence of the molar heat capacity which is shown in Figure 12.1. At 273.2 K the heat capacity decreases by 50% if the water freezes. If, on the other hand, the water is able to undercool, then the $C(T)$ curve exhibits no discontinuity. Indeed, as the temperature decreases to below 260 K, C rises increasingly sharply with decrease in temperature, so that the heat capacity difference between ice and water increases correspondingly. Similar divergence effects are observed for most other physical properties, e.g. density and compressibility. The origin of this rapidly increasing temperature sensitivity of most physical properties of undercooled water

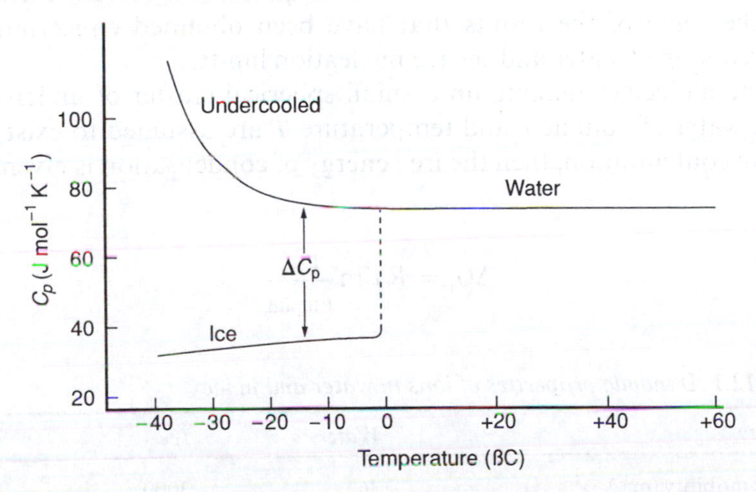

Figure 12.1 *Heat capacities of liquid water and ice as function of temperature* (Reproduced by permission of Cambridge University Press from F. Franks, *Biophysics and Biochemistry at Low Temperatures*, 1985)

will be discussed later in this chapter but it must be concluded that, at any given temperature, ice and undercooled water have dramatically different physical properties. Such differences are even more pronounced in the dynamic properties of dissolved ions and can amount to orders of magnitude, as shown in Table 12.1.

Homogeneous Nucleation of Ice in Undercooled Water

Any discussion of undercooling might use as its starting point the phenomenon of nucleation, the mechanism(s) by which a daughter phase (e.g. solid) can grow spontaneously in a mother phase, (e.g. liquid). The theory of nucleation was originally developed from a consideration of liquid droplet condensation in supersaturated vapours. As such, it is based on the stepwise addition of molecules to an existing embryo, until a certain critical size is reached, beyond which growth becomes spontaneous. This model appears to be reasonable when applied to the condensation of water vapour in the upper atmosphere to form clouds and give rise to precipitation. It is not obvious how such a model of stepwise aggregation can account for the growth of ice in undercooled water, bearing in mind the structural similarities of water and ice. Both substances exist as three-dimensional, hydrogen-bonded networks, making the definition of a boundary of an embryo, where condensation of water molecules occurs, problematical. Despite the above, and several other uncertainties, classical nucleation theory has provided a useful insight into the behaviour of liquid water below its equilibrium freezing point. It is therefore useful to describe some of the results that have been obtained concerning the undercooling of water and the ice nucleation limits.

If the molecules making up a small spherical cluster of undercooled liquid water of volume v and temperature T are assumed to exist in an ice-like configuration, then the free energy of condensation is given by

$$\Delta G_t = RT \ln \frac{p_{ice}}{p_{liquid}} \qquad (12.1)$$

Table 12.1 *Dynamic properties of ions in water and in ice*

Property	*Water*	*Ice*
Proton mobility ($m^2\ V^{-1}\ s^{-1}$)	36	3000
Li^+ mobility ($m^2\ V^{-1}\ s^{-1}$)	4	$<10^{-4}$
Li^+ self-diffusion coefficient ($m^2\ s^{-1}$)	0.24	8×10^{-7}

where p_{ice} and p_{liquid} are the sublimation pressure of ice and the vapour pressure of the undercooled liquid, respectively. Since at subzero temperatures ice is the stable phase, then $p_{ice} < p_{liquid}$, so that $\Delta G_t < 0$. Actually ΔG_t is the free energy change that accompanies the transfer of a mole of water from the undercooled liquid to the interior of the solid-like cluster. Equation (12.1) can be approximated to

$$\Delta G_t = \frac{-T\Delta H_t \Delta T}{T_f^{\,2}}$$

where ΔH_t is the latent heat of crystallisation and $\Delta T = (T_f - T)$. Opposing the growth of the cluster are the forces due to surface tension (σ) between the liquid and the cluster surface. For a cluster of radius r, the net free energy accompanying the liquid–solid condensation is

$$\Delta G_t = -4\pi r^2 \sigma - (4/3)\pi r^3 \, \Delta G_t \tag{12.2}$$

where σ is now the interfacial free energy of the cluster.

Condensation becomes spontaneous when $\Delta G_{l \to s} < 0$. For this to occur, the critical values are

$$r^* = -2/\Delta G_t \text{ and } \Delta G^*_{l \to s} = 16\pi r^3 (\Delta G_t)^2$$

The relationship between the various quantities is represented in Figure 12.2 for $T = 233$ K ($-40\,°$C).

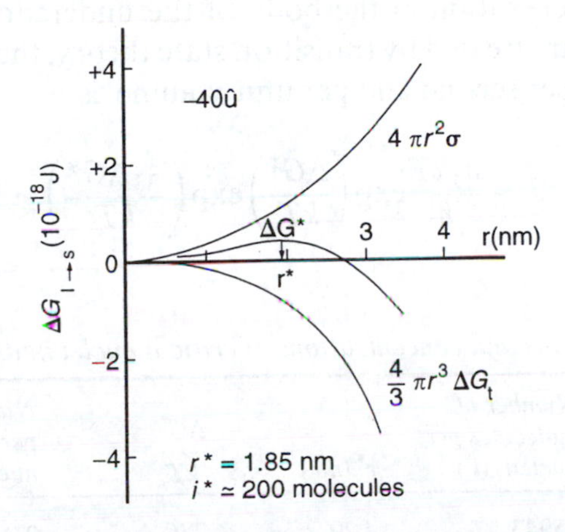

Figure 12.2 *Volume free energy of nucleation, $\Delta G_{l \to s}$, of water, as a function of the cluster radius at 233 K, according to equation 12.2. For details of the parameters used, see text*
(Reproduced by permission of Cambridge University Press from F. Franks, *Biophysics and Biochemistry at Low Temperatures*, 1985)

The number of critical clusters in a given volume of water at a given temperature is estimated by assuming a Boltzmann distribution:

$$n(r^*) = n_1 \exp -(\Delta G^*/kT) \qquad (12.3)$$

where n_1 is the number of molecules per unit volume in the liquid phase. The properties of water clusters and critical nuclei are computed in Table 12.2 as a function of temperature. Consider a temperature of 243 K: $r^* = 1.38$ nm and i^*, the number of molecules in a critical cluster, is 566. The number of active nuclei in a gram of water is thus very small, well below the limit of experimental determination. From the data in Table 12.2 it is possible to calculate the mass of water which is likely to contain one critical nucleus at a given temperature. Somewhere between 243 and 233 K a temperature is reached at which every gram of water is likely to contain one nucleus capable of initiating freezing. If the mass of water were now to be divided up into droplets of radius 10 nm, then it is possible to calculate the proportion of drops that are likely to contain a nucleus and will therefore freeze at that temperature. This is shown in Table 12.3 for several temperatures. Although the actual numbers given in this sample calculation are subject to a considerable degree of uncertainty, they nevertheless demonstrate that nucleation is extremely sensitive to changes in temperature and that the very large numbers involved in the calculations change rapidly, by many orders of magnitude, over narrow ranges of temperature.

It is also important to consider the kinetics of nucleation, that is, the rate of nucleus generation in the body of the undercooled fluid. If nucleation kinetics are treated by transition state theory, then J, the number of nuclei formed per second and per unit volume, is

$$J \sim \frac{n_1 kT}{h} \exp\left(\frac{\Delta G^{\ddagger}}{kT}\right) \exp\left(\frac{-\Delta G^*}{kT}\right) \qquad (12.4)$$

Table 12.2 *Dimensions and concentrations of critical nuclei in undercooled water*

Temperature (K)	*Number of molecules per nucleus (i^*)*	*r^* (nm)*	$\frac{\Delta G^*}{kT}$	*Number of nuclei per gram of water nuclei g^{-1}*
263	15943	4.20	776	2.3×10^{-315}
253	1944	2.08	190	1.5×10^{-60}
243	566	1.38	83	3.8×10^{-14}
233	234	1.03	45	6.3×10^{2}
223	122	0.83	29	7.1×10^{9}

Table 12.3 *Distribution of ice nuclei within dispersed water droplets (based on data in Table 12.2)*

Temperature (K)	*Radius of drop (nm)*	r^* *(nm)*	*Number of droplets per gram of water*	*Number of nuclei per gram of water*	*Proportion of droplets containing a nucleus*
263	10	4.9	2.4×10^{17}	2×10^{-433}	8.4×10^{-499}
253	10	2.26	2.4×10^{17}	8.2×10^{-75}	3.4×10^{-92}
	5	2.46	1.9×10^{18}	2.5×10^{-93}	1.3×10^{-111}
243	10	1.46	2.4×10^{17}	2.5×10^{-18}	1.1×10^{-35}
	5	1.55	1.9×10^{18}	1.4×10^{-23}	7.2×10^{-42}
233	10	1.08	2.4×10^{17}	6.3	2.6×10^{-17}
	5	1.14	1.9×10^{18}	0.029	1.5×10^{-20}

The second exponential term is the Boltzmann distribution of nuclei, as given by equation (12.3) and the first exponential is the kinetic term which depends on the mechanism by which the clusters are formed, i.e. by the stepwise addition of molecules diffusing to the cluster surface. $\Delta G^{\ddagger}$ is therefore taken to be the free energy of activation of self-diffusion. Since self-diffusion is related to reciprocal viscosity, the activation free energy for viscous flow is commonly used in equation (12.4). By insertion of the appropriate physical properties of water, the following orders of magnitude for the various quantities in equation (12.4) are estimated:

$$\frac{n_1 kT}{h} = 10^{41}\ \mathrm{m^{-3}\,s^{-1}}$$

and

$$\left(\frac{n_1 kT}{h}\right)\exp\left(\frac{-\Delta G^{\ddagger}}{kT}\right) = 10^{36}\ \mathrm{m^{-3}\,s^{-1}}$$

in the neighbourhood of $-40\,°C$. J should therefore increase rapidly with decreasing temperature. Near the nucleation threshold the rate changes by about a factor of 20 per degree, so that homogeneous nucleation is a well-defined event, which does not greatly depend on the rate of cooling.

By making various substitutions for ΔG^* in equation (12.4) it becomes possible to express J in a fairly simple form:

$$J = A \exp(B\tau) \tag{12.5}$$

where τ is a quantity which contains all the temperature terms in equation (12.4); it is given by $\tau = [T^3(\Delta T)^2]^{-1}$. A more useful way of expressing τ is by way of *reduced* temperatures, where $\theta = T/T_f$ and $\Delta\theta = (T_f - T)/T_f$.

This makes possible the comparison of liquids and solutions which freeze at different temperatures. Equation (12.5) then becomes

$$J = A\exp(B\tau_\theta)$$

where $\tau_\theta = [\theta^3(\Delta\theta)^2]^{-1}$. A plot of ln J against τ_θ should therefore be linear, and this has been confirmed experimentally over several order of magnitude in J.

In the form of small droplets, water can thus be undercooled to approximately −40 °C, corresponding to a reduced degree of undercooling $\Delta\theta = 0.14$. Water is not particularly proficient at undercooling but its ability to undercool can be considerably enhanced by high pressures. This is illustrated in Figure 12.3, showing an enlarged version of part of the ice phase diagram in Figure 4.4

Starting at −40 °C, corresponding to atmospheric pressure, T_h falls rapidly with increasing pressure, reaching −90 °C at a pressure corresponding to the ice I–ice III transition. For every degree depression of the freezing point, T_h decreases by approximately two degrees, although there does not exist an obvious formal relationship between freezing (thermodynamics) and nucleation (kinetics).

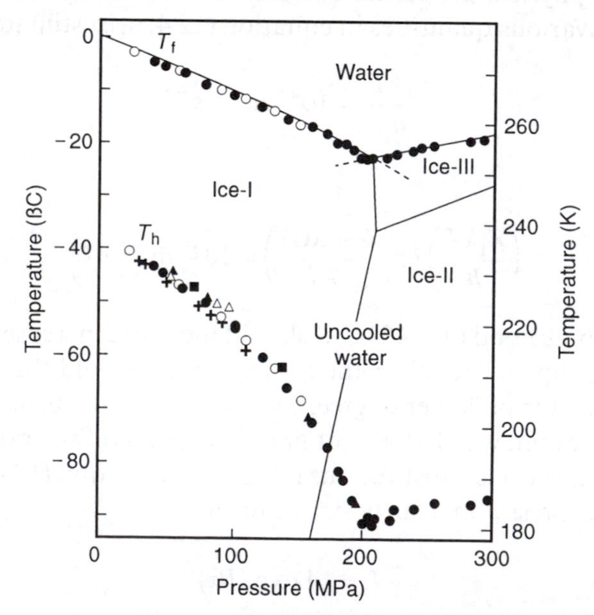

Figure 12.3 *Effect of pressure on the homogeneous nucleation of ice in undercooled water. Note how the nucleation curve mirrors the solid/liquid phase coexistence curve* (Reproduced by permission of Cambridge University Press from F. Franks, *Biophysics and Biochemistry at Low Temperatures*, 1985)

Nucleation of Ice by Particulate Matter

In practice, the homogeneous nucleation of water or ice is a very rare event; it probably takes place in the upper atmosphere where supersaturated water vapour condenses to rain, snow or hail. Much more common is the process whereby nucleation is catalysed by a solid or liquid substrate that allows groups of sorbed water molecules to take up configurations that are able to promote further condensation. This process is known as heterogeneous nucleation, but it must be emphasised that the actual nuclei so produced are clusters of water molecules identical to those considered above. The function of the foreign surface is that of a heterogeneous catalyst: it enhances the probability that a cluster of critical dimensions can form and persist for a sufficiently long time for further condensation to occur.

Much effort has gone into establishing what properties of the substrate are significant in rendering it active as a catalyst. Originally it was believed that a match of the crystal structure with that of ice was important, a hypothesis that resulted in the choice of silver iodide to promote nucleation of water and ice in supersaturated water vapour. The present consensus is that three factors determine the nucleating efficiency of a substrate: (i) it should have a small lattice mismatch with ice, (ii) it should have a low surface charge and (iii) it should possess a degree of hydrophobicity, i.e. water should not spread on it.

In recent years attention has focused on nucleating agents of biogenic origin. Certain microorganisms, insects and plants contain materials that possess high concentrations of very efficient ice nucleators. Indeed, heterogeneous nucleation is one of the mechanisms used by freeze-tolerant organisms to minimise the undercooling of their aqueous body fluids. Many, if not all, biological cells and tissues contain structures that are able to catalyse ice nucleation, although in some cases these structures are not of a high nucleating efficiency. Some structure (perhaps the internal face of the plasma membrane) contained in human erythrocytes promotes the freezing of intracellular water at a temperature which lies only 0.5 °C above that of T_h of an isotonic saline solution. Yeast cells, on the other hand, contain more efficient catalytic sites, which promote freezing at 10 °C above T_h.

The flowering plant *Lobelia telekii* provides a fascinating example of the protection afforded by efficient ice nucleation. Its natural habitat is on the slopes of Mount Kenya at temperatures that fluctuate daily within the range of −10 to +10 °C. The inflorescence of this plant contains a potent ice nucleating catalyst, which suppresses undercooling. When the temperature of the tissue fluids falls to the equilibrium freezing point,

$-0.5\,^{\circ}C$, freezing occurs and the latent heat released prevents further falls in the temperature of the plant during the night. The catalyst, believed to be of polysaccharide origin, has been shown to be active *in vitro*: it can substantially inhibit the undercooling of microdroplets of saline solution. Figure 12.4 shows *in vitro* freezing profiles of saline solution containing low amounts of polysaccharides, compared to the freezing behaviour of purified saline solution. It is seen that, whereas 50% of the saline solution drops freeze at $-16\,^{\circ}C$, droplets of fluid taken from the plant and also its polysaccharide fraction freeze close to $-10\,^{\circ}C$, with droplets of polygalacturonic acid solution freezing between these two extremes. Under *in vivo* conditions, bulk water in the plant freezes at $-0.5\,^{\circ}C$.

Heteronucleation does not necessarily require the presence of foreign matter in the bulk or at the surface of the undercooled liquid. It can also be induced by a sudden compression, the application of an electric field or irradiation. Indeed, the concept of nucleation was discovered by Fahrenheit in 1724, when he stumbled whilst carrying a flask filled with undercooled water and found that the water had frozen as a result of being shaken. The mechanisms whereby extraneous factors can influence the nucleating potential of an undercooled liquid are still subject to speculation.

Water Near its Undercooling Limit

Reference has already been made (see Figure 12.1) to the anomalous temperature dependence of the heat capacity of water as T_h is ap-

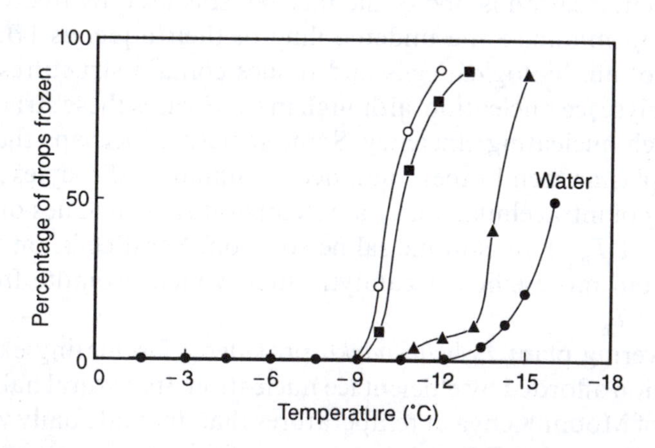

Figure 12.4 *Undercooling and droplet freezing of water (●), aqueous solution of polygalacturonic acid (▲), fluids extracted from the efflorescence of* Lobelia telekii *(○) and their polysaccharide fraction (■)*

proached. The same phenomenon of apparent divergence has also been observed with other physical properties (Y) which can then be expressed by a common relationship

$$Y = A_r\left(\frac{T}{T_s} - 1\right)^{-\gamma}$$

where T_s is the temperature of the singularity. The exponent γ is remarkably constant for a wide variety of physical properties, giving $T = 228$ K. Note that this temperature is several degrees below T_h, so that it cannot be reached experimentally at atmospheric pressure.

There is at present no general agreement yet about the origin of the eccentric temperature dependence of the physical properties of undercooled water as T_h is approached. It is not even certain that a singularity exists, because small angle X-ray scattering shows no abnormalities as T_s is approached.

Given that some form of anomalous behaviour is indicated, then two views currently exist about its origin: it might be due to a spinodal decomposition or to the existence of two critical points. Both hypotheses can rely on experimental as well as theoretical evidence. The divergence of the physical properties can be completely suppressed in the presence of some solutes, especially those that chemically resemble water, e.g. H_2O_2. On the other hand, the abnormalities become much more pronounced in solutions of hydrophobic-type solutes.

Glassy Water

If a liquid can be undercooled, then, in principle, it should be capable of vitrification and exhibit the phenomenon of a glass transition. The ease of vitrification depends largely on the rate of crystal growth within the undercooled liquid. In the case of water, the rate of ice growth is extremely high, thus making vitrification by fast cooling difficult. In addition, the physical properties that determine the cooling rate, i.e. the thermal conductivity, heat capacity and density, are unfavourable in the case of water and ice, which makes fast cooling a challenging experimental feat. Rates of $> 10^6$ K s^{-1} and liquid volumes not exceeding 10^{-13} m^3 are required to circumvent the nucleation and growth of ice. These demanding conditions have actually been achieved by quench cooling water films on electron microscope grids into a cryogenic fluid, such as liquid propane, or by spraying an aerosol onto a cold copper plate.

Amorphous solid water (ASW) can also be prepared by the condensation of water vapour on a metal plate, usually copper, at 77 K or by the low-temperature compression of ice-I above 10^3 MPa. When ASW is heated at atmospheric pressure to ca. 120 K, it is transformed into another type of ASW, with a reduction in the density. Glassy water can thus exist in two distinctly different phases, now described as low-density amorphous (LDA), and high-density amorphous (HDA). On heating, LDA undergoes a glass transition. The measured glass temperature T_g depends on the heating rate, but it lies in the range 124–136 K. Above T_g another transition, a crystallisation to cubic ice, is observed at ca. 150 K. There is as yet no agreement as to the actual nature of some of the phases; the whole field is still a subject of intense discussion and dispute. For obvious reasons, ASW is also a fertile field for computer simulation studies.

From a practical viewpoint, amorphous solid phases of aqueous solutions are of extreme importance in both ecology and various branches of technology. Some of these aspects will be explored in Chapter 13.

Superheated Water

The preparation of metastable water, whether by cooling or heating, requires a high degree of purity and the absence of materials that might act as catalysts for the return to the stable phase. By injecting small droplets of the purified liquid into a heated immiscible oil (benzyl benzoate), water has been successfully superheated to 552.7 K. Eventual boiling takes place explosively which raises the question whether such superheating events might cause hazards in some industrial processes. Compared to the importance of undercooling, however, the superheating of water is probably more of a scientific curiosity.

Chapter 13

Supersaturated and Solid Aqueous Solutions

Equilibrium and Metastable Aqueous Systems

Conventional physical chemistry texts are limited mainly to treatments of equilibrium systems and to processes that can be studied in real time and within the confines of the standard laws of chemical kinetics. As regards aqueous systems, liquid–vapour, liquid–liquid and solid–liquid phase equilibria are treated with reference to the Phase Rule. It is the purpose of this chapter to explore the behaviour of aqueous solutions outside the usual equilibrium phase coexistence boundaries. The emphasis is thus on supersaturated solutions which, by definition, are thermodynamically unstable. Their apparent stability is achieved by virtue of high energy barriers which severely retard the achievement of equilibrium states.

A striking example of such a system is provided by a mixture of oxygen and hydrogen gases at room temperature. The free energy balance predicts that, at 25 °C, the stable state is liquid water, but thermodynamics provides no information about the rate of conversion. Although thermodynamically classed as 'unstable', the gaseous mixture is remarkably stable kinetically, and no conversion can be measured in real time. Upon the addition of a catalyst, e.g. MnO_2, however, the gases react explosively to yield liquid water.

The concepts described in the previous chapter, which were developed for studies of unstable or metastable water, will therefore be used but will be extended to binary and some multicomponent mixtures. A distinction must thus be drawn between physical and/or chemical processes that *may* take place (thermodynamics) and those that actually do take place *within a measurable period of time* (kinetics). In practice, kinetics often appear to control solute nucleation and crystallisation from highly concentrated aqueous solutions. In such systems, supersaturation is therefore common.

The coupled phenomena of supersaturation and slow crystallisation are extensively utilised in practice and also have an important ecological dimension, to be discussed below.

Freeze Concentration

If a dilute aqueous salt solution, e.g. NaCl, is subjected to slow cooling, then after some initial undercooling, ice will be nucleated and crystallisation will proceed, following the liquidus curve, as described by the equilibrium phase diagram (Figure 13.1). Note that Figure 13.1 is not the conventional representation of a solid–liquid phase coexistence diagram, but it illustrates the process of freeze concentration more graphically. At the eutectic temperature of −21 °C, where the solution concentration has reached ca. 4 M, i.e. the saturation solubility at that temperature, equilibrium thermodynamics predict that $NaCl.2H_2O$ will precipitate. However, like ice crystallisation, salt precipitation also requires the growth and adequate lifetimes of viable nuclei. If nucleation and/or crystal growth are slow compared to the applied cooling rate, then the solution will become supersaturated in NaCl for an indefinite period. This is illustrated in Figure 13.2, where the concentration is drawn as a function of time, at a

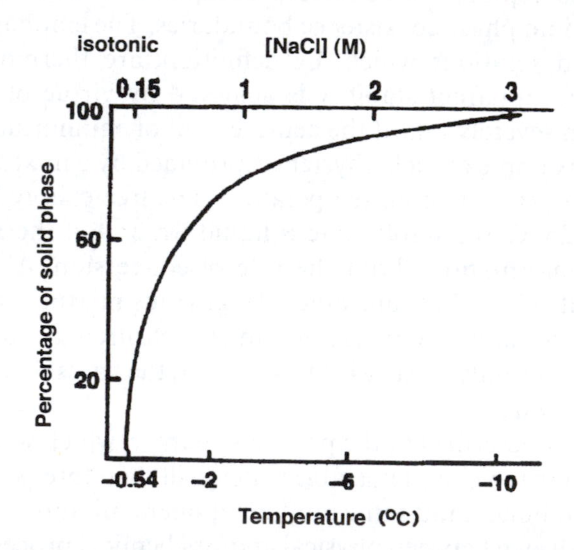

Figure 13.1 *Effect of cooling on the proportion of ice formed in an isotonic saline solution (0.15 M) after seeding with an ice crystal to prevent undercooling. Note that after cooling to −10 °C, the salt concentration of the residual liquid phase has risen to 3 M*
(Reproduced by permission of Cambridge University Press from F. Franks, *Biophysics and Biochemistry at Low Temperatures*, 1985)

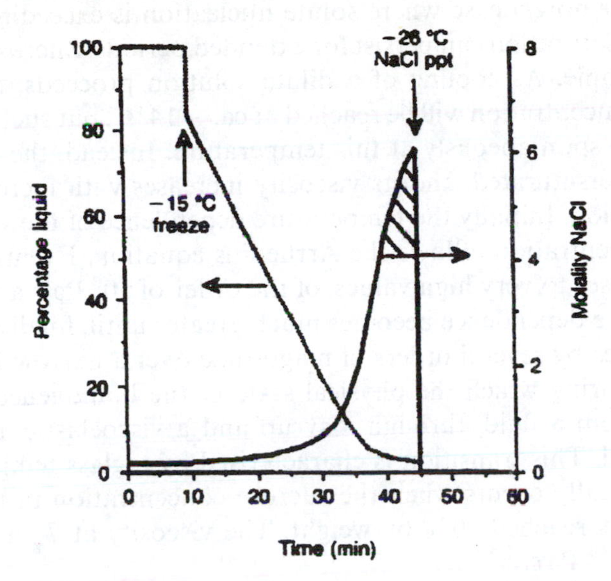

Figure 13.2 *Simulation of the typical freeze concentration course of an isotonic saline solution during cooling, as a function of time. Note the initial undercooling prior to ice nucleation at −15 °C. The shaded area represents the supersaturation range*

constant cooling rate. Since nucleation is a stochastic process, the degree of undercooling and supersaturation shown here can only serve as an example of what might occur. In this particular case, ice nucleation occurred after holding the solution at −15 °C for 10 min. Thereafter, the salt concentration increased with cooling, but the salt did not precipitate at the eutectic point. Instead, salt precipitation took place at −26 °C, when the concentration had reached 6 M. In other words, the supersaturated system regained equilibrium after ca. 10 min, where its final state then consisted of the equilibrium mixture of ice and $NaCl.2H_2O$ crystals. The above example illustrates the introduction of a time dimension (kinetics) into what is usually regarded as a thermodynamic equilibrium. Supersaturation thus takes the form of a transient excursion from equilibrium. The persistence of such an excursion may be anything from milliseconds to centuries, or even geological time-scales.

Although there is no formal theoretical connection between the osmotic melting point depression ΔT_m and the homogeneous nucleation temperature of ice T_h, it is found experimentally that a linear relationship exists between these two quantities: for each degree of melting point depression, T_h decreases by two degrees. The possible reasons for such a dependence are as yet unexplained, but the practical implications as regards technical undercooling are clear.

Consider now a case where solute nucleation is exceedingly slow, so that supersaturation can persist for extended periods. Sucrose serves as a good example. As cooling of a dilute solution proceeds, the notional eutectic concentration will be reached at ca. −14 °C, but sucrose does not precipitate spontaneously at this temperature. Instead, the solution becomes supersaturated, and its viscosity increases with increasing freeze concentration. Initially the temperature dependence of the viscosity at a given concentration follows the Arrhenius equation. Eventually, as the viscosity rises to very high values, of the order of 10^8 Pa s and above, its temperature dependence becomes much greater until, finally, the viscosity increases by several orders of magnitude over a narrow temperature interval, during which the physical state of the homogeneous solution changes from a fluid, through a syrup and a viscoelastic 'rubber' to a brittle solid. This transition is characterised by a glass temperature, T_g, which typically occurs when the sucrose concentration in the aqueous mixture has reached 80% by weight. The viscosity at T_g is then of the order of 10^{14} Pa s.

The aqueous glass is thus an undercooled liquid (noncrystalline) with a very high viscosity, equivalent to a flow rate of the order of 1–10 μm a^{-1}. Freezing comes practically to a halt at T_g, and the amorphous product is characterised by its glass temperature and composition. Figure 13.3 shows a typical solid/liquid *state diagram*, in this case for the binary water/sucrose system. This state diagram is not to be confused with a *phase* diagram, because, unlike the liquidus and solidus curves which meet at the eutectic point, the T_g/composition curve is *not* a phase coexistence curve but represents the locus of all compositions having the same viscosity, η_g. Because the crystal structure of sucrose is in the form of a complex, hydrogen-bonded three-dimensional network (akin to that of ice, but more complex), the nucleation of the sugar in real time, at a subzero temperature is most improbable. Instead, the solution continues to freeze and becomes progressively more supersaturated, thermodynamically unstable, but kinetically very stable.

Unlike the eutectic point (T_e), the point at which the liquidus curve intersects the glass curve, referred to in the literature as (T'_g, c'_g), is *not* an invariant point on a phase diagram, because freezing is incomplete, although it can no longer be detected on any practical time-scale. The twin phenomena of supersaturation and vitrification find interesting applications in food process technology and in the formulation of stable pharmaceutical products (see below). They are also believed to be implicated in natural drought and freezing resistance exhibited by a wide variety of living species.

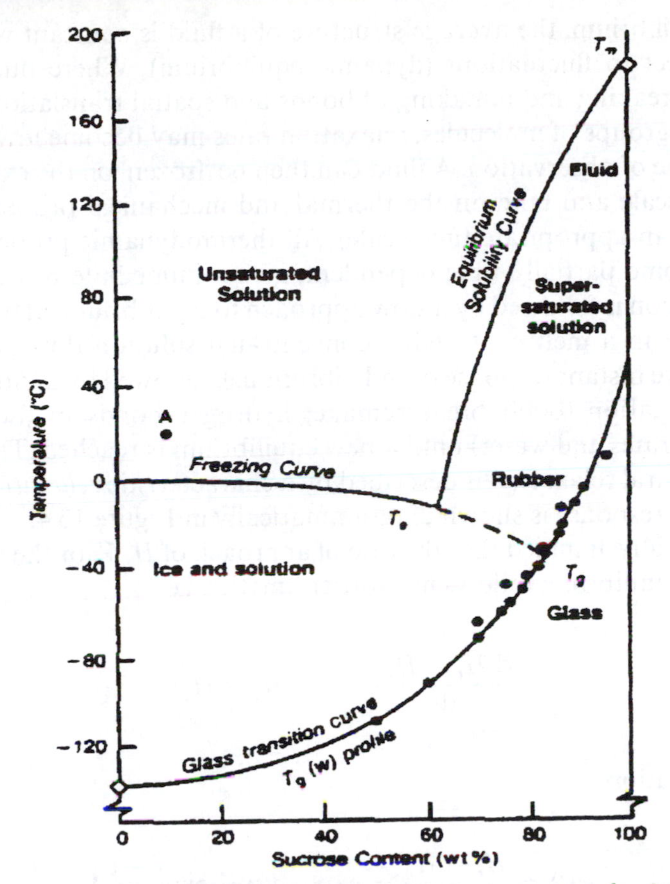

Figure 13.3 *Solid–liquid state diagram of the sucrose–water system, showing the glass transition/concentration profile, the equilibrium solubility curve (solidus), running between the melting point and the eutectic temperature* T_e, *and the large region of metastability (supersaturation) beyond* T_e *(broken line). Point A corresponds to a typical dilute solution at room temperature that is to be subjected to drying*
(Reproduced by permission of Cambridge University Press from F. Franks, *Biophysics and Biochemistry at Low Temperatures*, 1985)

Properties of Supersaturated Aqueous Solutions

Supersaturated aqueous solutions, especially those of polyhydroxy compounds (PHCs), tend to be highly viscous, even at ambient temperatures. Their thermomechanical properties (viscous flow, gelling, non-Newtonian rheology, etc.) have interested scientists and technologists for many years. Relaxation rates in such systems are governed by relatively slow rearrangements within their complex hydrogen-bonded configurations. For a proper description of such slow processes we need to consider the evolution in time of the system's equilibrium properties.

At equilibrium, the average structure of a fluid is constant with time, but subject to fluctuations (dynamic equilibrium). Where fluctuations involve breaking and remaking of bonds and spatial translation of molecules or groups of molecules, relaxation rates may become low, relative to the time of observation. A fluid can then be 'frozen' on the experimental time-scale and take on the thermal and mechanical properties of a solid (on an appropriate time-scale). All thermodynamic properties will then become partially time-dependent, i.e. an immediate response to a perturbation is followed by a slow approach to equilibrium. Step changes in T or P in a melt or a highly concentrated solution thus produce a crystal-like instant response (bond vibrations), followed by a 'slow' structural relaxation (bond break/remake, hydrogen-bonds in the case of carbohydrates and water) until a new equilibrium is reached .The rate of the structural relaxation is described by a characteristic *relaxation time* τ. This dual response is shown diagrammatically in Figure 13.4.

It is usually assumed that the rate of approach of H, V, or the configuration to equilibrium follows first-order kinetics, i.e.

$$\frac{\mathrm{d}(H_t - H_e)}{\mathrm{d}t} = -k(H_t - H_e)$$

On integration:

$$\phi(t) = \frac{H_t - H_e}{H_0 - H_e} = \exp(-kt) = \exp\left(\frac{-t}{\tau}\right)$$

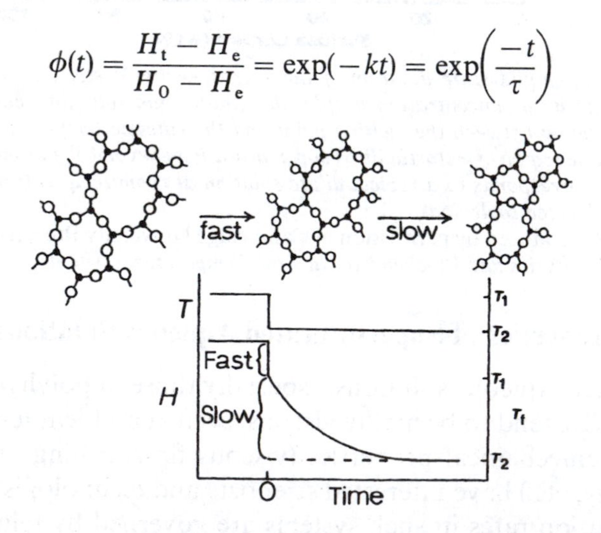

Figure 13.4 *Schematic plot of enthalpy versus time during isothermal structural relaxation following a step change in temperature*
(Reproduced by permission of the Mineralogical Society of America from C.T. Moynihan, *Rev. Mineral.*, 1995, **32**, 1)

where the relaxation function $\phi(t)$ changes from 1 (at $t = 0$) to 0 (at equilibrium).

More refined treatments take account of multiple relaxation processes (stretched relaxation) by including a 'distribution parameter' β in the relaxation function:

$$\phi(t) = \exp\left(\frac{-t}{\tau}\right)^{\beta}$$

The temperature dependence of τ is usually estimated by means of the Arrhenius expression:

$$\tau = \tau_0 \exp\left(\frac{\Delta H^*}{RT}\right)$$

It is found that the energy of activation, ΔH^*, is equal to the energy of activation of viscous flow.

The relaxation responses of a fluid to stepwise cooling and heating are shown diagrammatically in Figure 13.5. A constant temperature drop ΔT is applied, after which the system is allowed to relax for a time period Δt. It is seen that, initially, the response is fast enough for the system to regain equilibrium during the time period. As the temperature is decreased further, a stage will be reached where the relaxation becomes too slow for the attainment of complete equilibrium. This is the region of the glass transition, i.e. $\Delta t \approx \tau$. At even lower temperatures, the system will respond to further perturbations as a solid. Figure 13.5 also demonstrates

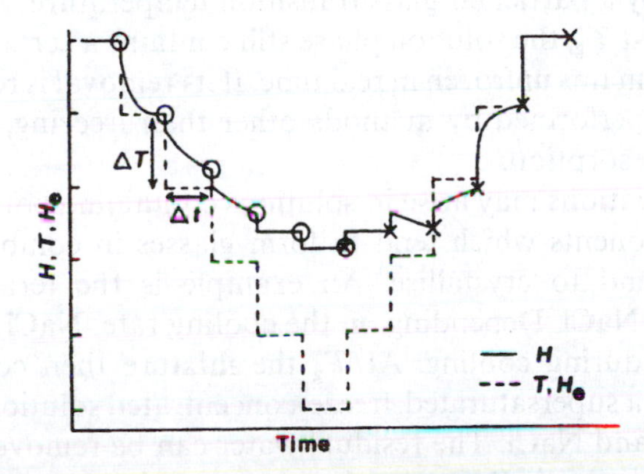

Figure 13.5 *Relaxation responses of a glass to stepwise cooling and heating* (Reproduced by permission of the Mineralogical Society of America from C.T. Moynihan, *Rev. Mineral.*, 1995, **32**, 1)

how and why in this region the $H(T)$ curve for cooling differs from the corresponding heating curve. It is also apparent that the observed transition region is cooling/heating rate dependent.

Amorphous States and Freezing Behaviour

The conversion of a dilute, unsaturated aqueous solution into a substantially dry amorphous solid by water removal involves several phase transitions, e.g. liquid–solid and solid–gas, but water and solutes do not necessarily undergo the same phase transitions, or at the same time. As the temperature is lowered, the solution will undercool (i.e. cool below the equilibrium freezing temperature) before freezing commences. During freezing, the solution remains in equilibrium with the solid phase (ice). The solutes remain in a concentrated, residual liquid phase. As the temperature is lowered further, the solution will become increasingly concentrated. In principle, variations in the cooling rate can affect the morphology of the ice formed during the freeze-concentration stage.

For multicomponent systems, solute crystallisation may or may not occur during freezing; it may also remain incomplete, depending on the solution composition and the cooling rate. Unless very low cooling rates are employed, and the frozen solution is allowed to anneal, some degree of supersaturation is common, with ice continuing to crystallise and the solutes remaining in the residual solution. At a characteristic temperature and composition freezing stops, at least at a measurable rate. The remaining mixture, excluding the crystalline ice phase, then undergoes the glass transition. This transition from a supersaturated, freeze-concentrated aqueous solution of very high viscosity to a brittle solid is operationally characterised by a particular glass transition temperature, T'_g. As shown in Figure 13.3, at T'_g, the solution phase still contains a certain amount of water, which remains unfrozen in real time. If its removal is required, then this has to be performed by methods other than freezing, e.g. heating, diffusion and desorption.

Complex situations may arise in solutions containing both sugars and salts, i.e. components which tend to form glasses in combination with those which tend to crystallise. An example is the ternary mixture water–sucrose–NaCl. Depending on the cooling rate, NaCl may or may not crystallise during cooling. At T'_g the mixture then comprises ice, coexisting with a supersaturated, freeze-concentrated solution containing water, sucrose and NaCl. The residual water can be removed by careful heating or exposure to P_2O_5, leaving an amorphous solid solution of NaCl in sucrose. Further heating to a temperature just above T_g causes the slow crystallisation of well-formed cubic crystals of $NaCl.2H_2O$. This is graphically shown in Figure 13.6, a scanning electron micrograph. Salt crystallisation is seen to take place preferentially at the high-energy sites, provided by stress cracks in the glass. The construction of a full state diagram for such a ternary mixture, containing several crystalline and amorphous phases, and giving rise to one ternary and seven binary eutectic and two peritectic mixtures, is a major experimental feat and can only be achieved by the combination of several experimental techniques. Such information is, however, extremely useful for the development of stable pharmaceutical preparations and the correct setting of freeze-drying process parameters.

During recent years, concentrated PHC solutions have been subjected to detailed studies and previously unknown and 'unsuspected' crystalline hydrates have been discovered of raffinose, trehalose and mannitol. They may be considered as metastable forms that can only be prepared under particular experimental conditions. Mannitol hydrate is of particular interest because no hydrate had previously been known of this polyol. It is apparently formed when a freeze concentrated solution is slowly dried at a temperature below T'_g. Figure 13.7 shows its X-ray diffractogram. Several of the lines, marked with asterisks, are not found in the diffraction

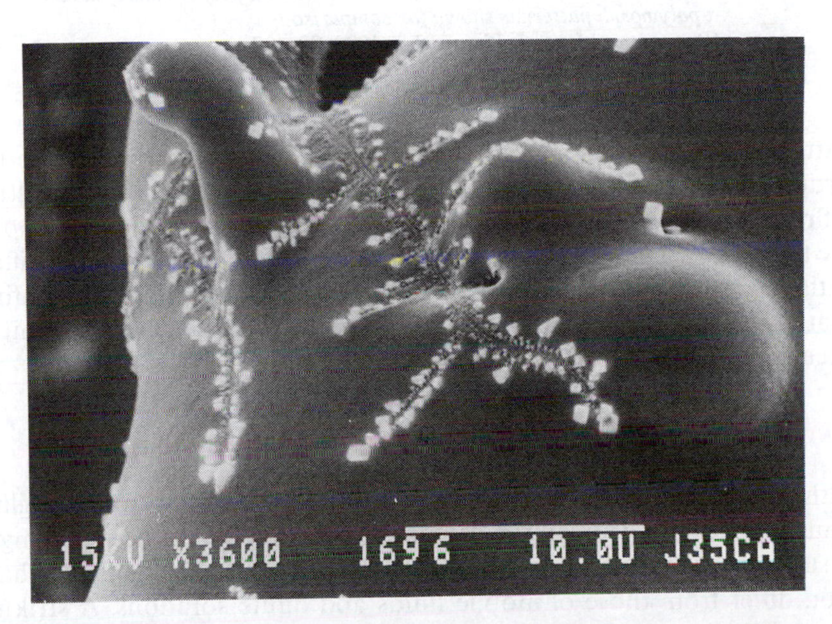

Figure 13.6 *Scanning electron micrograph of a freeze-dried solution containing 10% sucrose + NaCl with a mass ratio 5:1. The dried material contained 2% residual water. After a short exposure to a temperature above T_g, cubic $NaCl.2H_2O$ crystals of micrometer dimensions have grown in the amorphous substrate ('solid solvent'). Scale bar = 10 μm*

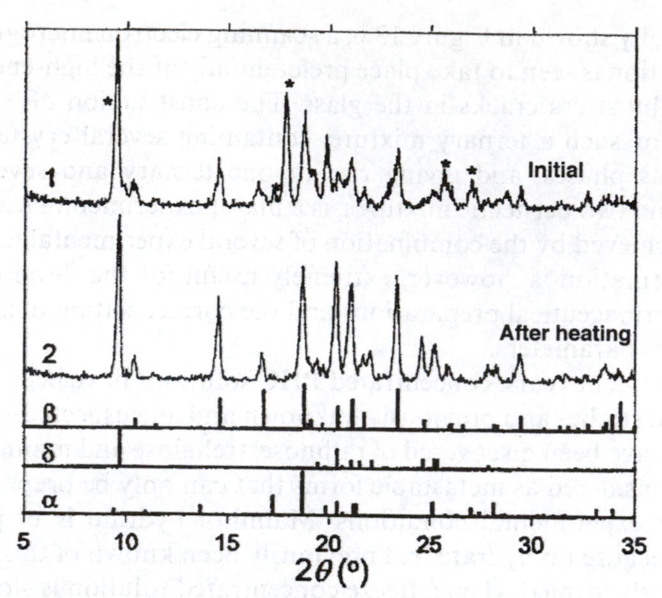

Figure 13.7 *X-ray powder diffraction patterns of a freeze-concentrated mannitol solution (1) and of the same solution after heating (2). Diffractogram 1 contains peaks (*) that are not attributable to any known mannitol polymorph. All the peaks in diffractogram 2 are attributable to the δ and β polymorphs of mannitol. The α polymorph pattern is shown for comparison*
(Reproduced from *J. Pharm. Sci.*, 1999, **88**, 197)

patterns of the known mannitol polymorphs. They disappear when the hydrate is heated for 30 min at 70 °C (lower diffractogram). The resulting diffraction pattern then contains only the characteristic peaks of two of the three known anhydrous mannitol polymorphs. Corresponding differential scanning calorimetry and thermogravimetry heating scans confirm that a stepwise weight loss of 6% (one mole H_2O per mole mannitol) occurs near 50 °C.

Dynamics in Supersaturated and Vitrified Solutions

Although there is as yet no consensus about the exact nature of the glass transition, experimental observations indicate that the properties of highly supersaturated, viscous solutions, and especially their dynamic behaviour, differ from those of mobile fluids and dilute solutions. A striking example is provided in Figure 13.8 which shows the rotational relaxation time of sucrose, determined by nuclear magnetic relaxation measurements, relative to that of water in mixtures of increasing sucrose concentration. As expected, in dilute solutions, the motions of the two species are

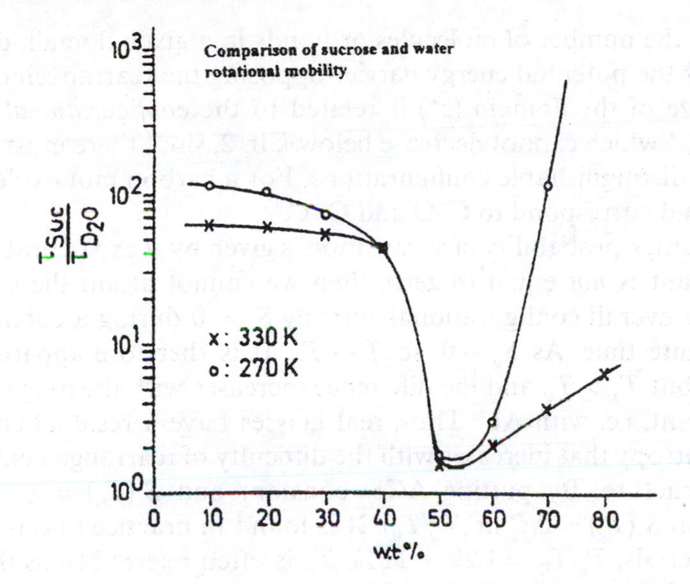

Figure 13.8 *Rotational mobility of sucrose τ_{suc} relative to that of water τ_{D_2O} as a function of sucrose concentration*
(Reproduced From D. Gurlich, Ph.D Thesis, University of Regensburg, 1991)

uncoupled, i.e. water is able to rotate much faster than sucrose. With increasing concentration, however, the rotation times of the two species become coupled, only to decouple once again at even higher concentrations, as the glass transition is approached. Relaxation times of the water molecules, even in the glassy sucrose solution phase, remain remarkably low, suggesting that water retains its liquid character. This type of dynamic decoupling is not taken into account in classical kinetic theories of solutions, but it may have far-reaching practical consequences.

Theory of Structural Relaxation by Cooperative Motions

Below T_g, molecular relaxation times are too long for equilibrium to be established within an experimental time-scale, because of a reduction in the number of accessible configurations, expressed in terms of a configurational entropy. The probability of a configurational rearrangement as a function of temperature can be written as

$$W(T) = A \exp\left(\frac{-z\,\Delta U}{kT}\right)$$

Here, z is the number of molecules or bonds in a given domain domain, and ΔU is the potential energy barrier opposing the rearrangement. The critical size of the domain (z^*) is related to the *configurational critical entropy* S_c^* which cannot decrease below $k \ln 2$, since there must exist at least two distinguishable configurations. For a carbon monoxide lattice these would correspond to C–O and O–C.

The average probability of a transition is given by $A \exp(\text{-const}/TS_c)$. If the constant is *not* equal to zero, then we cannot attain the situation where the overall configurational entropy $S_c = 0$ during a cooling process of finite time. As $S_c \rightarrow 0$, so $T \rightarrow T_0$. It is therefore apparent that $T_g \neq T_0$, but $T_g > T_0$ and the difference increases with the magnitude of the constant, i.e. with ΔU. Thus, real glasses have a residual configurational entropy that increases with the difficulty of rearrangements within the structure. By putting $\Delta C =$ constant and $S_c(T_0) = 0$, then by integration $S_c(T_g) = \Delta C \ln (T_g/T_0)$. It is found in practice that for many 'real' materials, $T_g/T_0 = 1.29 \pm 11\%$. T_0 is often referred to as the temperature of zero mobility, or the Kauzmann temperature. There is as yet not enough information to test such relationships for their applicability to aqueous mixtures.

From a semi-empirical treatment of relaxation processes it has been inferred that $W(T)$ is proportional to τ^{-1}, and thus to diffusion, viscous flow and conductivity. If $[(T - T_0)/T_0] \ll 1$, then it can be shown that $W(T) = A \exp [B/(T - T_0)]$. This is one form of the Vogel–Tammann–Fulcher (VTF) equation (see below) which adequately describes diffusion and viscous flow, but not in the neighbourhood of T_g.

Kinetics in Amorphous Solids

The VTF equation can also be written as

$$\tau = \tau_0 \exp\left[\frac{BT_0}{(T - T_0)}\right]$$

where B is related to the energy of activation ΔH^* of the relaxation process, and τ_0 refers to the high-temperature limit of unrestricted mobility and is often taken to lie 17 orders of magnitude below the experimental τ value. When $T_0 = 0$, the VTF equation reduces to the Arrhenius equation, with B proportional to ΔH^*. When $T_0 > 0$, ΔH^* becomes temperature-dependent.

The so-called Williams–Landel–Ferry (WLF) equation provides a special case of the VTF equation:

$$\eta = \eta_0 \exp\left[\frac{C_1(T - T_g)}{C_2 + (T - T_g)}\right]$$

where C_1 and C_2 are coefficients that describe the temperature-dependence of the particular relaxation process at temperatures above the reference temperature, here taken as T_g. In practice, the temperature-dependence of the relaxation time constant τ is usually best described by the VTF equation.

An examination of reduced viscosity/temperature plots, several of which are shown in Figure 13.9, reveals that some materials (SiO_2) display Arrhenius behaviour, i.e. ln (η/η_g) varies in a linear manner with T_g/T. For other materials the viscosity falls off much more steeply than is predicted by the Arrhenius equation. Such differences have been rationalised in terms of a so-called 'fragility' parameter as a means of classifying fluids. In strong fluids, the structure (mainly covalently bonded) is maintained above T_g, whereas in fragile fluids the structure is rapidly disrupted above T_g, probably reflecting their weakly hydrogen-bonded structures. According to this classification, water, carbohydrates and their (solid) aqueous solutions are fragile, whereas native proteins are strong materials, a surprising result that still awaits a convincing explanation.

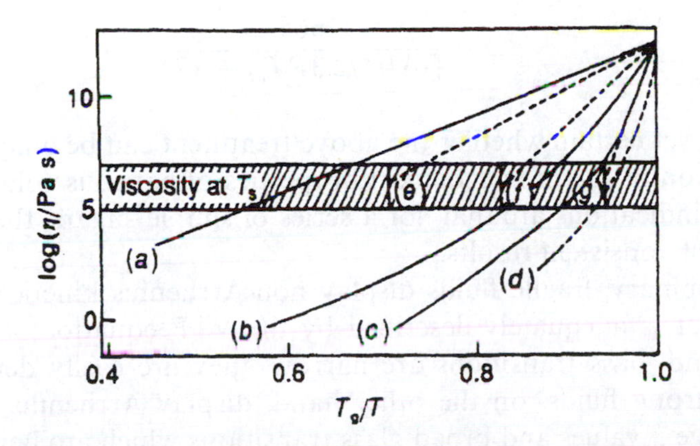

Figure 13.9 *Viscosities of glass-forming materials, scaled to the glass transition and illustrating the phenomenon of strong (a) and fragile (d) fluids. (a) silica, (b) glycerol, (c) 80% sucrose solution, (d) 100% quenched sucrose, (e) ovalbumin solution; (f) and (g) maximally freeze-concentrated solutions of glycerol and sucrose, respectively. The shaded area represents the viscosity range where viscous flow/collapse can be detected experimentally in real time. Viscous flow in strong fluids follows Arrhenius kinetics; the VTF equation best describes the flow kinetics of fragile fluids. For further details see text*

The fragility parameter δ is defined as follows:

$$\delta = \frac{(\Delta US^*/k)}{(T_g \Delta C_{g \to l})}$$

and $T_g/T_0 = f(\delta)$. The VTF equation can then be expressed in terms of the relaxation time, referred to T_0:

$$\tau = \tau_0 \exp\left(\frac{\delta T_0}{T - T_0}\right)$$

On the assumption that 17 orders of magnitude separate the relaxation times at T_g and at some common high temperature limit, corresponding to complete molecular flexibility, then a linear relationship is obtained between T_g/T_0 and δ:

$$\frac{T_g}{T_0} = 1 + 0.025\delta$$

With the aid of some of the above equations and data from 'real' materials (oxides, silicates, etc.), it has been possible to obtain δ values which are remarkably constant for given classes of materials:

$$\delta = \frac{665}{[(\Delta H^*/2.3RT_g) - 17]}$$

It is not yet certain whether the above treatment can be adapted for the calculation of thermomechanical properties of aqueous solid solutions, but the indications are that, for a series of simple sugars, the treatment yields self-consistent results.

In summary, fragile fluids display non-Arrhenius kinetics, and their behaviour is adequately described by the VTF equation; δ values are small, and glass transitions are narrow; they are easily detectable by DSC. Strong fluids, on the other hand, display Arrhenius behaviour, with large δ values and broad glass transitions which are hardly detectable by DSC.

Significance of Glassy States in Drying Process Technology

The glass transition profile shown in Figure 13.2 illustrates the ability of small amounts of dissolved water to depress T_g of sucrose. Water is thus able to reduce the mechanical strength of the dry amorphous solid (glass) and exerts a softening influence, commonly referred to as plasticisation. Water has been termed the *ubiquitous plasticiser* for all organic materials. The semi-empirical relationship between T_g and the composition of a material is given by the Gordon-Taylor equation which, for a binary mixture, can be written as:

$$T_g = \frac{wT_2 + k(1 - w)T_1}{w + k(1 - w)}$$

where w is the weight fraction of one of the components, T_1 and T_2 are the glass transition temperatures of the two pure components and k is a fitting parameter that does, however, possess some physical significance. The very low glass transition of water (135 K) accounts for its pronounced plasticisation capability.

The loss of mechanical strength (decrease in viscosity) of sucrose, resulting from even a low amount of sorbed water, also illustrated in Figure 13.9, is an example of a phenomenon which is of major importance technologically, as well as ecologically. Many chemically and/or microbiologically labile materials achieve an apparent stability after having being dried from aqueous solution. For example, food and pharmaceutical processing industries are much concerned with product shelf lives and thus have come to rely on the thermomechanical properties of aqueous glasses (solid aqueous solutions). Just as relaxation rates decrease by orders of magnitude at the glass transition, so chemical reaction rates are affected in a similar manner. Although such a stabilisation effect is attractive in principle, it must be borne in mind that the drying of a dilute aqueous solution (ca. 5–10% solid content) to a glass, containing ca. 2–3% residual water, requires the removal of >99% of the water, either by freezing or by evaporation, without causing chemical damage to any labile species present in the supersaturated phase, either in solution or suspension.

In practice, molecules which possess a limited stability in aqueous solution, e.g. native proteins, can be formulated with so-called excipients before drying, the function of the excipient being to protect the labile molecule against deleterious reactions during drying, and thereafter, and also to provide an amorphous matrix that will inhibit undesirable long-term chemical changes. In the case of pharmaceutical preparations, the excipients must fulfil some quite demanding conditions: they must be compatible with the bioproduct and of an acceptable solubility. They must not react with the biomolecule, even in supersaturated solution, or undergo physical changes, such as phase separation by crystallisation.

Finally, they should have high glass transition temperatures, enabling a dried product to be stored for extended periods at ambient, or even elevated temperatures. For obvious reasons, pharmaceutical processes and products are also subject to stringent safety demands, and the list of 'permitted' excipients is short. Carbohydrates are particularly valuable in this respect, because they fulfil the above conditions, although with a few remarkable exceptions: of all the lower PHCs, mannitol and, to some extent, lactose are prone to crystallise from aqueous solution, even during freezing, rendering them of limited value as pharmaceutical protectants and glass formers.

Aqueous glass technology is a relatively new discipline, although studies of 'real' amorphous materials (metal alloys, synthetic polymers, silicates, oxides, etc.) go back many years, and materials science has established itself as an important branch of chemistry and physics. It has only recently been recognised that the thermophysical properties of water-sensitive materials, in particular natural products, also underpin important branches of product and process technology. Thus, it is well known that chemical and microbiological stability and safety of labile products can be achieved by drying, and that drying can be performed by removing water either as a solid (by freezing) or as a vapour (by boiling). What has only recently come to be appreciated is that successful drying protocols and the preparation of stable products depend on a knowledge of processes that may occur when a dilute solution becomes supersaturated and eventually undergoes a glass transition. In other words, the solution, initially thermodynamically stable, ends up kinetically stable, but during the transition it has to pass through a region of intermediate concentration where it becomes highly vulnerable to physical and/or chemical changes, most of which are irreversible and deleterious. This is illustrated in Figure 13.10, where the different routes that can be taken to remove water are also indicated.

Conventional wisdom has it that, because of the low temperatures used, freezing and frozen storage or freeze-drying are to be preferred over drying processes at ambient or elevated temperatures. This is, however, a misconception, because it does not take into account that freezing is in fact synonymous with drying (see Figure 13.1). It is also commonly believed that chemical reaction rates decrease during freezing, i.e. they follow Arrhenius kinetics. Actually reaction rates increase dramatically during freezing, because the effects of freeze concentration far outweigh the retarding effects of reduced temperatures. Some enzyme-catalysed reactions display rate enhancements of several orders of magnitude during freezing. Other problems that may be encountered during freezing include the possibility of major pH changes in buffered solutions, where

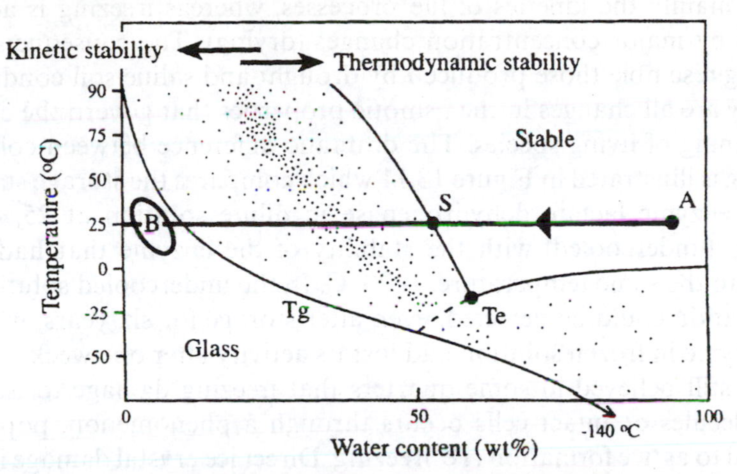

Figure 13.10 *Generalised temperature/composition stability map, showing the domains of thermodynamic stability (unsaturated solution) and kinetic stability (supersaturated solution below its glass temperatures). Water removal can be performed by freezing, followed by sublimation of ice, or by direct evaporation. The stippled area indicates physical and/or chemical vulnerability, with the density of stippling corresponding to the degree of instability*

one of the buffer salts precipitates at its eutectic point, whereas the other component remains in the supersaturated solution.

The above discussion is designed to provide evidence that some of the 'rules' which students of physical chemistry are taught are not only irrelevant but are based on assumptions that have no place in 'real world' chemical and biochemical processing. It is unfortunate that most research into aqueous mixtures, is even now still limited to dilute solutions, and that structural and dynamic properties of supersaturated systems, especially those close to the glass transition, remain largely unexplored.

The design of efficient and successful drying processes, whether by freezing or by evaporation, for chemically labile, water-sensitive materials are now seen to depend sensitively on the thermophysical and thermomechanical behaviour of supersaturated solutions during water removal, and thereafter. The knowledge gained from studies of amorphous solids can be readily applied to the trivial process of sugar candy manufacture and the not so trivial process of manufacturing stable biopharmaceuticals.

Natural Freezing and Drought Resistance

Cold and freezing are the most serious threats to many forms of life on land. A clear distinction must be drawn between the two effects: cold affects mainly the kinetics of life processes, whereas freezing is accompanied by major concentration changes (drying). The consequences of freezing resemble those produced by drought and saline soil conditions, i.e. they are all changes in the osmotic properties that govern the correct functioning of living species. The dramatic difference between cold and freezing is illustrated in Figure 13.11 which compares the storage stability of the enzyme lactate dehydrogenase in dilute solution at 25, 4 and −20 °C (undercooled) with the stability of the enzyme that had been frozen to the same temperature, −20 °C. In the undercooled solution no inactivation could be detected, even after storage for six years, whereas the enzyme in frozen solution had lost its activity after one week.

It is still believed in some quarters that freezing damage to isolated biomolecules or intact cells occurs through a phenomenon, popularly referred to as 'ice formation', i.e. freezing. Direct ice crystal damage is thus invoked as the primary cause of injury. While such a process may contribute to vascular tissue damage, it had already been demonstrated many years ago (see Figure 13.12), that haemolytic damage to isolated red blood cells, associated with freezing, could be exactly simulated by the addition of salt to the isotonic medium, *at room temperature.*

The sensitive dependence of various life forms on water management is graphically demonstrated in Figure 13.13. The left-hand scales represent

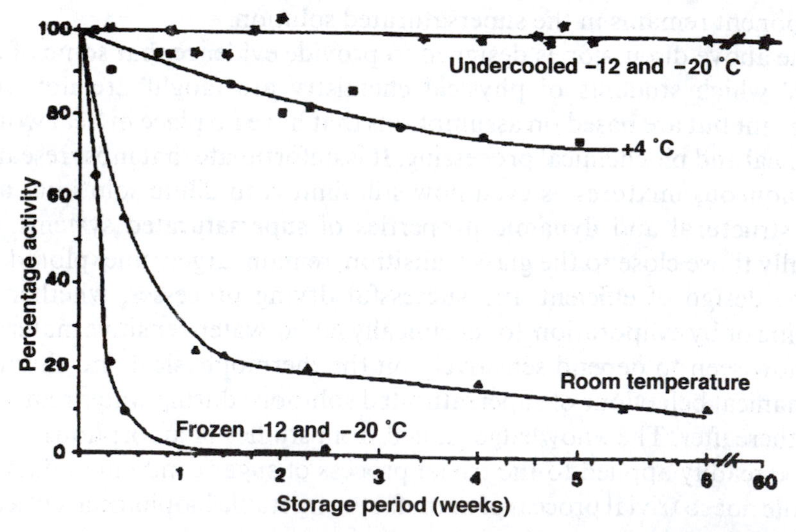

Figure 13.11 *Storage stability, monitored by enzyme activity, of lactate dehydrogenase in dilute solution at different temperatures. Note particularly the striking difference between storage in the frozen state and in the undercooled state, at the same temperature*

(Reprinted from *Process Biochem.*, **22**. Hatley, Franks and Mathias, 171. © 1987 with permission from Elsevier Science)

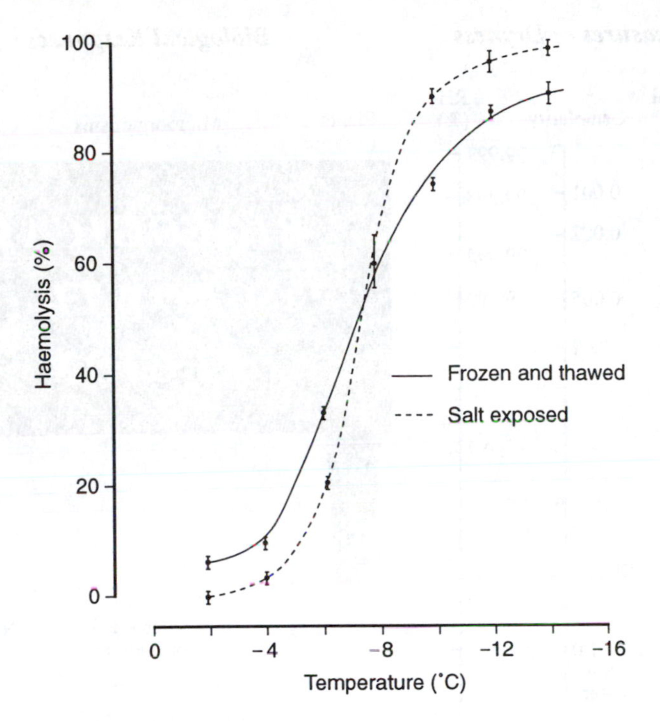

Figure 13.12 *Haemolysis of erythrocytes in phosphate-buffered saline solution on freezing and on exposure to a salt concentration equivalent to that produced during freeze concentration*

various measures of 'dryness', while the right-hand side lists the observed effects of water removal on several species. The amazing result is surely that optimum biological functioning of most species is limited to an extremely narrow range of osmotic conditions, corresponding to relative humidities between 99.9 and 99.999%. Any conditions lying outside this range are perceived as a physiological water stress. Viability then rapidly decreases with the intensity of the environmental stress. Organisms that are completely adapted to conditions outside this range, e.g. halophilic, thermophilic or psychrophilic species, are classified as extremophiles. Their biochemical machinery, such as protein and/or lipid compositions, is adapted to function 'normally' in their particular environment. They, in turn, would sense 'normal' conditions as stress.

Many nonextremophilic species are, however, able to survive even quite severe osmotic stress conditions for extended periods. The chemical adaptation mechanism can be triggered *in vitro* by the transient imposition of a slight stress, e.g. hypertonic salt concentrations, the shortening of the photoperiod, low temperature, reduction of the relative

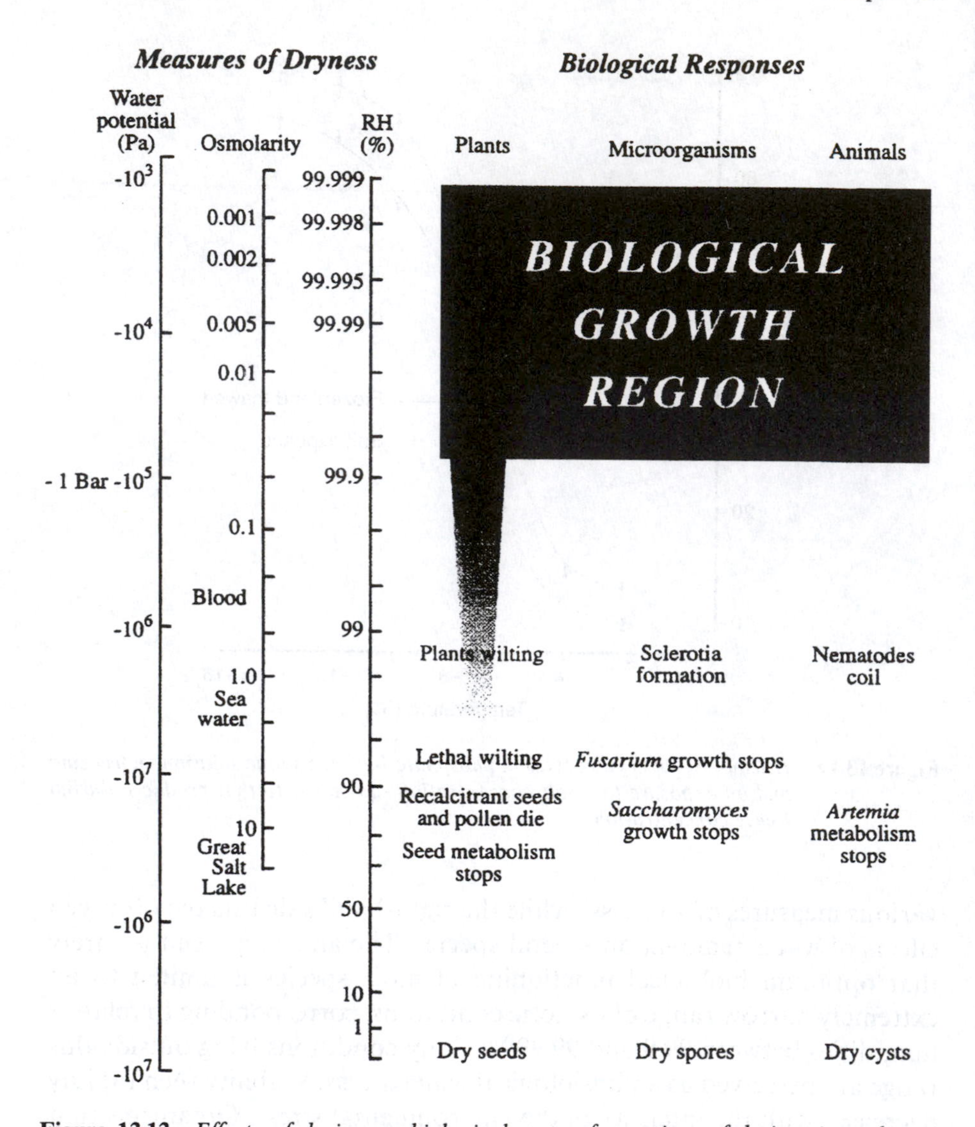

Figure 13.13 *Effects of drying on biological state of a variety of desiccation tolerant organisms*
(Adapted from *Membranes, Metabolism and Dry Organisms*, ed. A.C. Leopold, Cornell University Press, 1986)

humidity, reduction in nutrient levels, etc. Such imposed stresses trigger major metabolic changes in the course of which the organism's starch reserves are converted into low-molecular-weight PHCs. An alternative chemical change consists of the accumulation of free amino acids in the body fluids. Such amino acids may either be those that occur in proteins, e.g. glutamic acid or proline, or they may also be non-protein amino acids, e.g. ectoine (a condensation compound of lysine), strombine (the

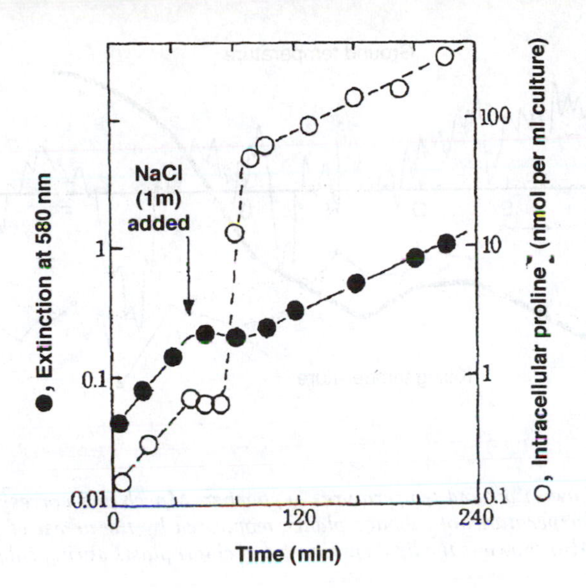

Figure 13.14 *Osmoregulation by* Bacillus subtilis *in a growing culture, after being subjected to a salt shock (1 M NaCl). The left-hand ordinate is a measure of protein synthesis*
(Reproduced with the permission of The Royal Society from G.W. Gould and J.C. Measures, *Phil. Trans. R. Soc.*, 1977, **B278**, 151)

amino acid analogue of succinic acid) or a variety of betaines (alkyl substituted glycine).

The period required for complete adaptation (hardening) to severe future environmental conditions depends on the species and the eventual stress intensity; it may last from several minutes, in the case of some microorganisms, to several months (autumn) for higher plants. Some examples of adaptation mechanisms* are given in Figures 13.14–13.16. Figure 13.14 traces the accumulation of proline and the kinetics of protein synthesis in *Bacillus subtilis*, after suffering a hypertonic NaCl shock. Over a very short period free proline levels increase by two orders of magnitude, while protein synthesis is effectively halted. After adaptation is complete, protein synthesis recommences, although at a reduced level. Figure 13.15 shows the relationship between the recorded ground temperatures during autumn and winter and the corresponding temperature at which photosynthesis of cabbage plants ceases. The hardening and dehardening of the plants follows closely the fluctuations in ground temperature. The change in the lipid content of chloroplasts during winter hardening is also shown; unlike photosynthesis, it is a more

* The American term for this process is 'acclimation'.

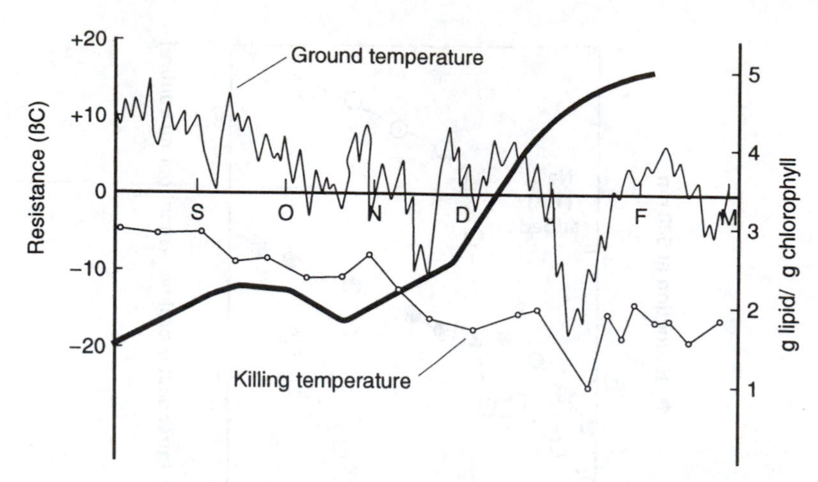

Figure 13.15 *Lowest ground temperatures September–March and corresponding 'killing' temperatures of cabbage plants, monitored by the arrest of photosynthesis. Also shown is the lipid enrichment of chloroplasts during cold adaptation*

long-term effect and does not correlate with the short-term temperature fluctuations. Figure 13.16 summarises the winter profiles of glycerol, sorbitol and glycogen in the freeze tolerant larvae of the gall fly *Eurosta solidaginis.* Here again, high levels of PHCs serve to protect the organism against the damages of freeze concentration. The connection with the properties of supersaturated PHC solutions, discussed earlier, is evident. In fact, in many cases of remarkable resistance to desiccation, e.g. dry seeds and bacterial spores, the probable occurrence of vitrification of body fluids loaded with PHCs cannot be discounted.

The whole subject of drought and freezing tolerance in plants has received a major boost with the advent of genetic modification. It was soon realised that the development of resistant crops was a multi-gene problem, even after it had already been established which chemical species are involved as protectants against the major osmotic stresses: drought, freezing and salinity. Carbohydrates and various amino acids, some of them chemically modified and not all of them found in proteins, are the most prolific. The plant growth regulating hormone abscisic acid (ABA) also appears to be implicated in desiccation tolerance.

The role of the disaccharide trehalose in promoting desiccation tolerance has been a subject of intense investigation and publicity over the past 12 years, although convincing evidence for any unique properties possessed by this sugar is hard to come by. Many species can develop seasonal desiccation tolerance with the help of other protectants. In most recent developments, *E. coli* genes, which encode trehalose biosynthetic

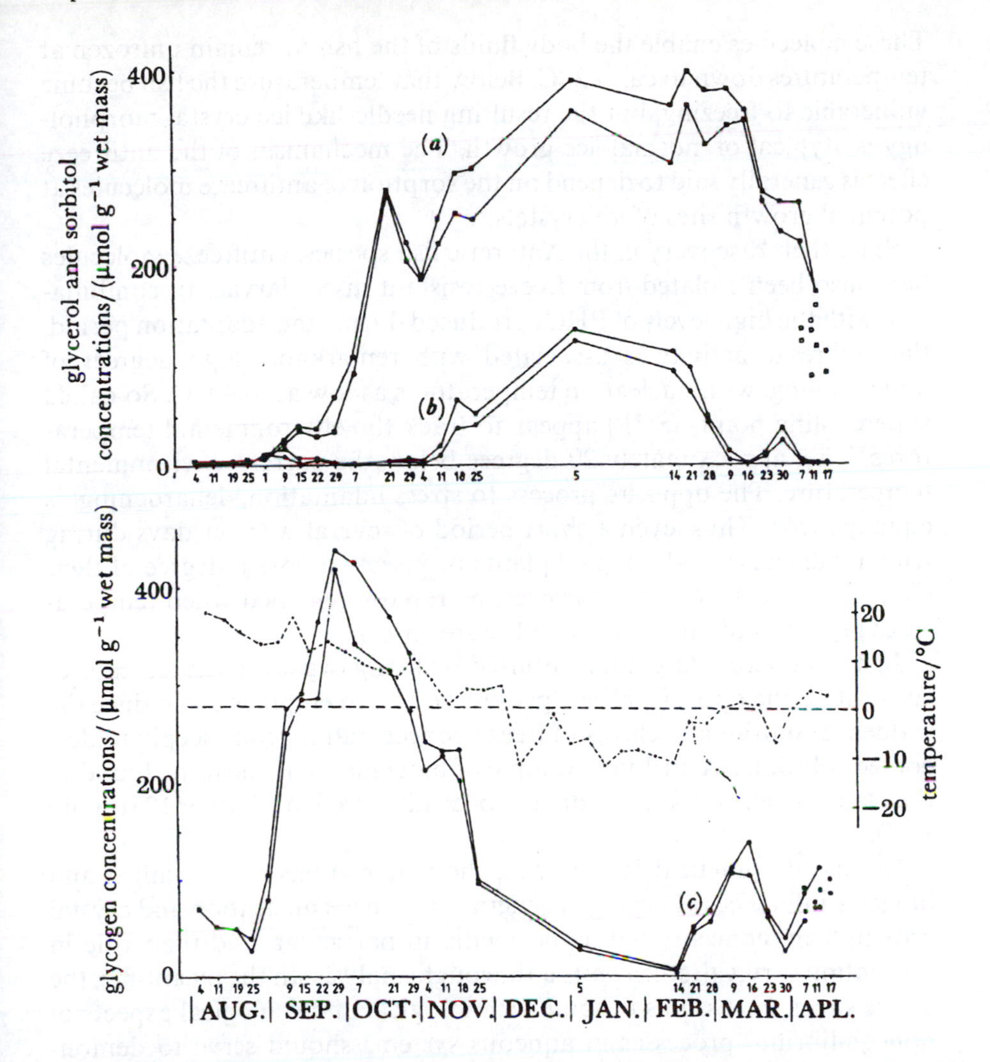

Figure 13.16 *Winter profiles of glycerol* (a), *sorbitol* (b) *and glycogen* (c) *in the freeze tolerant larvae of the gall fly* Eurosta solidaginis; *also plotted are the mean weekly environmental temperatures*
(Reproduced with permission of The Royal Society from *Phil. Trans. R. Soc.*, 1990, **B326**, 635)

enzymes, have been successfully expressed in human fibroblasts, rendering these cells resistant to drying, the first example of osmotic stress acclimatisation by mammalian cells.

Finally, mention must be made of an alternative mechanism of *in vivo* survival: freeze avoidance. The phenomenon of long-term undercooling was first studied in Antarctic fish species and led to the discovery of so-called antifreeze molecules, usually of peptidic or glycopeptidic origin.

These molecules enable the body fluids of the fish to remain unfrozen at temperatures down to ca. $-2\,^{\circ}C$. Below that temperature the fish become vulnerable to freezing, but the resulting needle-like ice crystal morphology is atypical of 'normal' ice growth. The mechanism of the antifreeze effect is generally said to depend on the sorption of antifreeze molecules at potential growth sites of ice crystals.

Since their discovery in the Antarctic fish species, antifreeze molecules have also been isolated from freeze-resistant insect larvae. In combination with the high levels of PHCs produced during the adaptation period, the antifreeze activity is associated with remarkably high degrees of undercooling, with nucleation temperatures as low as $-40\,^{\circ}C$. So-called supercooling points (SCP) appear to track the environmental temperatures, lying approximately 20 degrees below the current environmental temperature. The opposite process to stress adaptation, dehardening, is equally rapid. Thus, even a short period of several warmer days during winter will cause cold-adapted plants or insects to lose a degree of their cold tolerance which can, however, be rapidly regained when temperatures begin to fall once again (see Figure 13.15).

Just as undercooling can be utilised by living organisms as a means of survival, so the opposite effect, 'pro-freeze' is also employed to reduce the sudden, and dramatic effects of freeze concentration from deeply undercooled solutions. A striking example of biogenic ice nucleators, found in the plant *Lobelia telekii*, has already been discussed in Chapter 12 (Figure 12.4).

Despite its practical importance, the whole subject of chemical and biogenic influences of drying in general, and on ice nucleation and crystal growth from concentrated liquid media in particular, and their role in promoting survival, still awaits a thorough analysis. In the meantime, the above short discussions of some technological and ecological aspects of nonequilibrium processes in aqueous systems should serve to demonstrate the important role played by water management both *in vivo* and *in vitro*.

Chapter 14

Water Availability, Usage and Quality

Water as an Essential Resource

Although not directly related to chemistry, several important aspects of water in our ecosystem touch on the fringes of the natural sciences and deserve some discussion. They are interlinked and provide a direct path from the exact sciences of physics and chemistry, via the life and earth sciences, to technological, economic and political considerations.

Apart from its universal importance as the elixir of most forms of life, water is also finding an increasing role in industry. With the general recognition of oil as a depleting resource, but also for environmental reasons, water is gradually replacing organic solvents in many chemical processes. This transition has presented chemical engineers with new challenges regarding rigorous purity standards and processing techniques.

Water Availability

The evaporation/precipitation cycle is of vast magnitude and ensures that the earth's water resources are recycled 37 times annually. The total quantity of water reaching the earth every year in the form of precipitation could cover the area of the globe to a depth of 0.5 m. However, to talk about an average annual 0.5 m of precipitation is misleading, because the actual distribution over the earth's surface is quite non-uniform and irregular in time as well as in space. Most of the precipitated water returns directly to the oceans; the remainder falls on land as a result of one of three processes involving the precipitation from warm, moisture-laden air:

(1) *Cyclonic*. The warm air mass, either stationary or in horizontal motion, encounters a cold air mass.

(2) *Convective.* An air mass, receiving heat and water vapour at the earth's surface, cools as it rises vertically.
(3) *In mountain regions.* A moving warm air mass is forced to rise and lose heat when it encounters a mountain barrier.

Meteoric water falls irregularly with respect to geographical location, with coastal areas receiving more. In the Northern Hemisphere the prevailing winds are westerly, so that the western slopes of mountain ranges generally receive adequate precipitation. As the air mass moves down the leeward sides it is again warmed and precipitation is decreased, resulting in a 'rain shadow' which, in extreme cases, can give rise to desert regions.

Of the average total rainfall, about 70% evaporates; the remainder appears as liquid water on or below the land surface. Some water, not included in the average rainfall figure, evaporates in the air between the clouds and the land surface. The remaining evaporation losses are of two forms: direct evaporation from wet surfaces and transpiration through plants from their leaves and stems. The precipitation adhering to trees and other vegetation (intercepted water) accounts for a large proportion of the total evaporation.

The 30% of the water not directly returned to the atmosphere by evaporation is called runoff and constitutes a region's potentially available water supply. Actually, the proportion of the earth's total freshwater resources which participates in the hydrologic cycle does not exceed 0.003%, the remainder being locked up in the Antarctic ice cap. Several feasibility studies are on record for the towing of icebergs to regions, e.g. in Australia, that are short of water. Although such a scheme has not yet been put into operation, there are no major technical barriers and the economics compare favourably with those of desalination.

Some of the runoff eventually collects in surface streams, but the larger proportion percolates underground where permeable soil is present. Gravity forces water downward through alluvial deposits, porous rock, gravel or sand. Ultimately, it may either reappear at the surface in artesian springs or seeps or as discharges below sea level into the oceans. The larger part of groundwater discharge reaches surface streams and keeps them flowing even during dry periods. Streams and rivers flow into the oceans which eventually receive over 90% of the total runoff.

Groundwater is of increasing importance as a supply source. Surprisingly, less than 3% of the world's available fresh water occurs in streams and lakes, but because of large, expensive surface water projects one tends to think of surface water as the most important potential source. Also, groundwater hydrology, as a relatively new science, developed slowly, although modern geological principles had already been formulated near the end of the eighteenth century. Shallow wells have been in use for many centuries, but early concepts of the origin of groundwater were curious mixtures of superstition and faulty deduction. In recent years the science of groundwater hydrology has advanced significantly, not only as a result of extensive hydrological data gathering but also from the explosion of research in botany, chemistry, physics and meteorology.

The resources allocated to groundwater hydrology vary considerably from one country to another. For example, the United Kingdom is not one of the most advanced countries in this respect; this became very evident during the freak drought of 1976. Irrigation engineers, on a visit to the UK from Israel, found it inconceivable that a country with very large *potential* water resources could suffer such acute water shortage just as a result of a few weeks without rain, nor could they understand how a situation could ever arise in the UK that would require such draconian measures as hosepipe bans, let alone rationing by standpipes.

Underground water flow comprises an estimated 33–40% of the average total runoff, but this average is greatly exceeded in some areas. Not all the groundwater is currently available at what is considered an economical cost; hence 80% of present supplies are drawn from surface sources. Nevertheless, five states of the United States (Arkansas, Arizona, Mississippi, New Mexico and South Dakota) depend on groundwater for over half of their total water supplies. Eight others, among them California and Texas, use groundwater for 25–50% of their supply. Surprisingly, some regions – such as Ontario Province in Canada – that are rich in lakes and rivers, nevertheless draw their domestic and industrial supplies exclusively from groundwater reservoirs, mainly for economic reasons. This can lead to severe problems during periods of rapid increases in population and industrial activity.

Groundwater is of crucial importance in semi-arid countries where large parts of the population depend on agriculture. Thus, in India more than 90% of the total water usage is accounted for by irrigation. Rainfall is highly seasonal and erratic, and successful agriculture is impossible without extensive irrigation. The irrigated area has been almost doubled since the inception of water resource planning about 50 years ago, but even now it only amounts to 30% of the gross cultivated area. Groundwater contributes 40% to total irrigation. While rainfall is the major source of groundwater, infiltration and storage of water depends mainly on the geological formations. These two factors provide a fair indication of the occurrence and distribution of groundwater resources across the country. The most important geological feature is the degree of compaction or consolidation. Figure 14.1 represents a hydrological map of India,

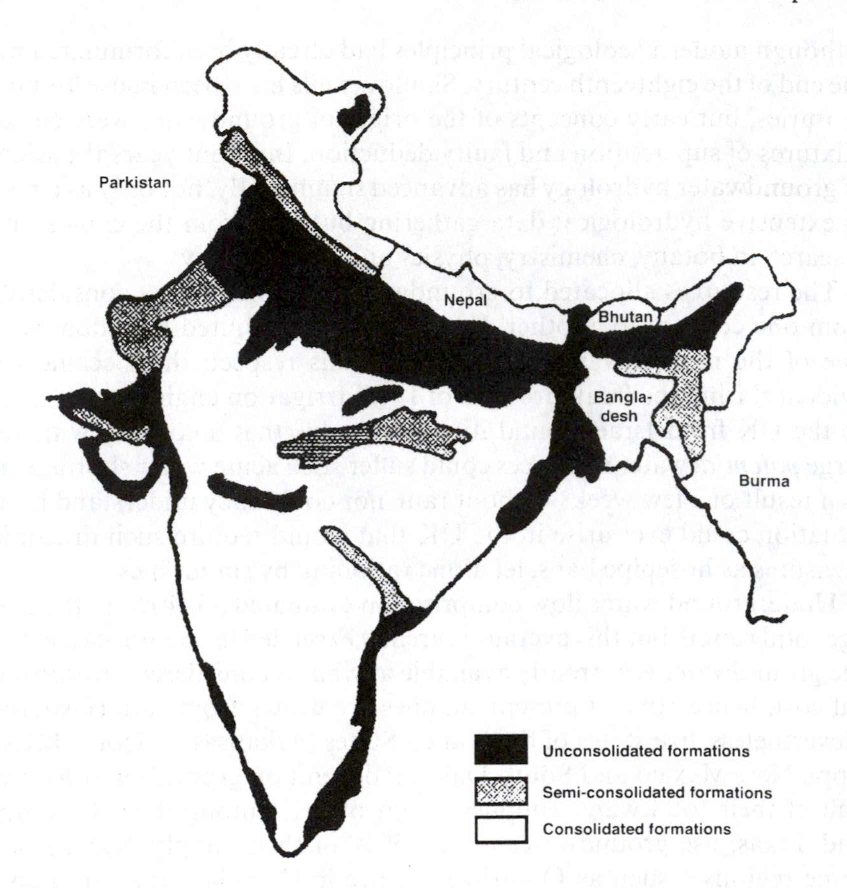

Figure 14.1 *Hydrological map of India, showing areas of abundant and scarce groundwater supplies*

showing that the major unconsolidated (porous) formulations correspond to the alluvial plains built up by the large river systems of the Indus, Ganges and Brahmaputra. These areas are also comparatively rich in rainfall. They account for 30% of the total land area but contain 50–60% of the usable underground water resources.

Nearly two-thirds of the landmass consists of hard rock formations, which contain no pore spaces and hold only limited amounts of water in fracture zones. Five per cent of the landmass consists of semi-consolidated formations, most of it in the Northwest which is, however, an arid region, receiving little rain. The groundwater is of uneven quality and partly mineralised. A sandstone formation has been discovered in this region which holds a 7000 year-old water reservoir, amounting to some 2.5 million ha m, but without possibility of recharge. Similar types of vast underground reservoirs have also been located in other desert areas, e.g. the Sinai Peninsula and Libya.

Most of the groundwater development in India takes place through simple wells of different types, which are cheap enough to be afforded by farmers on an individual basis, but a special public development programme was initiated for better utilisation of the irrigation potential. Coupled with this was an investigation programme for the continuous monitoring of water levels and changes in water quality.

The capricious incidence of rainfall is well demonstrated by the statistics, where such exist. One of the most arid locations, the Wadi Halfa in the Sudan, has a mean annual rainfall of 2.5 mm. Yet during the two wettest months on record 30 and 15 mm of rain fell. These periods of heavy rainfall occurred during a July and an October respectively. This type of pattern is not at all uncharacteristic for locations such as the Sahara, Saudi Arabia or the Australian Outback. The full exploitation of the considerable resources represented by such rains benefits from integrated networks, combining international meteorology with the observations and experience of desert nomads.

The maintenance and quality of groundwater reservoirs is also related to land-use and treatment practices. Developed countries show that semi-arid conditions need not be a bar to agricultural exploitation. Thus, the land and climate of arid and semi-arid Australia are comparable with those of Asia and Africa, but socio-economic factors are very different. For the past 120 years the grazeable part of arid and semi-arid Australia has been used for grazing sheep and cattle. Developments of the associated wool and meat industries have been a trial-and-error procedure with over-optimistic estimates of the productivity and resilience of the natural resources. The present level of exploitation may still be debatable, but over the past 40 years integrated research with long-term objectives has been carried out, and the results are apparent.

Similarly, in the United States, land-treatment programmes were started in the 1930s to restore the quality of the land and control runoff and erosion. In one test area, grazing control alone reduced runoff by 30% and sediment yield by 35%. Conversion of vegetation from shrubs to grass, while not affecting the water yield, reduced erosion by 14 times. Other effective practices include mechanical land treatment and floodwater spreading on valley floors. On the other side of the balance sheet all practices designed to increase plant cover, induce infiltration or control flows themselves require water. Therefore such practices must be judged with regard to their effect on the net water yield.

The intricate interrelationships between meteorology, geology and agriculture are also being explored by computer modelling techniques, but our understanding of the place/weather/crop system is still incomplete. For instance, the dependence of crop growth on solar irradiation is quite well established, but this is hardly the case for the relations between environmental factors and respiration, leaf growth and death, and the partition of dry matter.

Many statistical and experimental models, with weather variables as inputs, have been developed and are being used in the field. They are not yet very precise, but they do describe limits of biological yields as a function of the weather. Solar irradiation, and so probably potential evaporation, is more uniform over a period of years than is rainfall. In the humid tropics water stress is hardly a limiting factor in plant growth, but under more arid conditions a partial failure of rain may result in total crop failure; yield is therefore not necessarily a continuous function of rainfall. Experience indicates that, within the semi-arid tropics, a seasonal total rainfall of about 80% of the mean is probably as vital as the total quantity.

The International Water Management Institute (IWMI), with headquarters in Sri Lanka, has launched a major programme to improve and guide irrigation policies, especially in the Developing World. Over the past three decades increases in food production have been achieved by expanding the irrigated area and increasing crop yields per unit of water demand, mainly by the development of new varieties. The scale of increases in water usage and demand is best illustrated by some figures. Thus, of the 40 000 dams, all but 5000 have been built since 1950. At the same time the exploitation of groundwater has also increased, in some places to an alarming extent. In some semi-arid regions of China and India, some of the world's major breadbaskets, water tables have been falling by 1–3 metres annually. The overdraft of groundwater, equivalent to 600 km^3, is presently disguising the seriousness of future water shortages.

Many models have been developed in efforts to match water supply to demand and thus to estimate the water balance in different parts of the world. These models are not always realistic; for instance, the considerable amount of water taken up by trees is frequently forgotten in such calculations. In an effort to assist governments, IWMI has produced what is probably the most realistic model in efforts to estimate different scenarios of water supply and demand for the year 2025. Considering the assessment for 45 countries representing 83% of the world's population produces a reasonably optimistic picture for the ability of these countries to achieve increases in their food production. It will be recalled that the major food shortages and rising prices projected in studies in 1960 never actually happened, because remedial action was taken.

The IWMI model predicts that 2.7 billion people, including a considerable number of those living in India and China, will live in regions of severe water shortages. Water shortages may also lead to conflicts in the Middle East and Sub-Saharan Africa (see Chapter 15) where poverty is already extreme. Other, more affluent regions will, however, also be affected, because expanding demands are currently draining some of the world's large rivers, e.g. the Nile, theYellow River, the Colorado and the Rio Grande. Three major methods for increasing more agricultural output per unit of utilisable water are being suggested:

- Increase the potentially utilisable water resources through development of irrigation infrastructure and storage facilities,
- Consume more of the developed water resources by recycling or better storage,
- Produce more output per unit of water employed by improvements on crop varieties, agronomic practices and water management.

The situation is most acute in Asia, where the irrigated area almost doubled during the period 1965–1995, with a corresponding increase in the area of arable land of only 10%. In other parts of the world the increases were less dramatic. In Europe the area of arable land actually fell during the period, although the irrigated area increased by 30%.

The economics of world food production change constantly. Today 5% of the world's farmers produce half of the world's food. In some countries water scarcity will make it more economical to import cereal grain, rather than engage in large-scale engineering projects, designed to increase water supplies. This practice has been called 'trade in virtual water.' Thus, during the mid-1990s, approximately 100 million tons of wheat exports corresponded to an annual 'water trade' of at least 100 km^3. In the Far East, where the important crop is rice, 50% of the total crop is grown without recourse to irrigation, mainly in the regions of large river deltas. Thus, Viet Nam has become the world's second largest rice exporter.

IWMI is reasonably optimistic that the Doomsday scenario of a coming water crisis does not need to become reality, although major investments and policy changes will be required by many nations.

Supplementation of Water Resources by Climatic Modification – a Case Study

Reference has already been made to the various methods, which might be employed in certain geographical locations to increase the available supplies of fresh water. It is instructive to discuss the various ramifications of one of these methods in some detail. Climatic modification at first sight appears an attractive proposition, because any water generated does not result from a saving but constitutes a net increase in the supply. Moreover, this water will, by virtue of its origin, be of a high quality. Climatic modification is not yet being used to any great extent, not least because we lack the knowledge required to control precipitation with any precision.

In 1971 the US National Science Foundation commissioned a study to explore the feasibility of applying technology assessment concepts to planned weather modification projects. This study was based on a proposal to augment the flow of the Colorado River by cloud seeding in the Upper Colorado River Basin during winter, thus enhancing the accumulation of snow in the mountains. The overall assessment took into consideration the following factors:

(1) the cost/effectiveness of snowpack augmentation, based on the then current state of the art;
(2) an estimate of improvements in the cost/benefit ratio over the next five years, arising from improved technology;
(3) the ecological impact on the area;
(4) the public attitudes to the project;
(5) the legal consequences and potential public policy changes;
(6) comparison of snowpack augmentation with other means of alleviating water problems in the area and beyond.

Figure 14.2 shows the schematic approach adopted in the technological evaluation of the project. The findings of the study were published in 1974 and illustrate strikingly the interplay of technical, economic, sociological and juridical factors, which would need to be taken into consideration in arriving at public policy decisions.

Technically, the method of snow-cloud seeding by silver iodide provides an inexpensive way of increasing the water supplies. The method is illustrated in Figure 14.3. Compared to virgin snow, equivalent to 1.75 million ha m, an additional 0.44 million ha m could be produced at a cost which compares very favourably with most other water supplies. Cloud seeding would increase the snowfall at altitudes above 2500 m by 20–25%.

The environmental, economic and social impact is likely to be negative, but moderate. In fact, it would be no worse than the conditions normally encountered after a particularly severe winter. The environmental consequences would include increased danger of avalanches and floods. At a

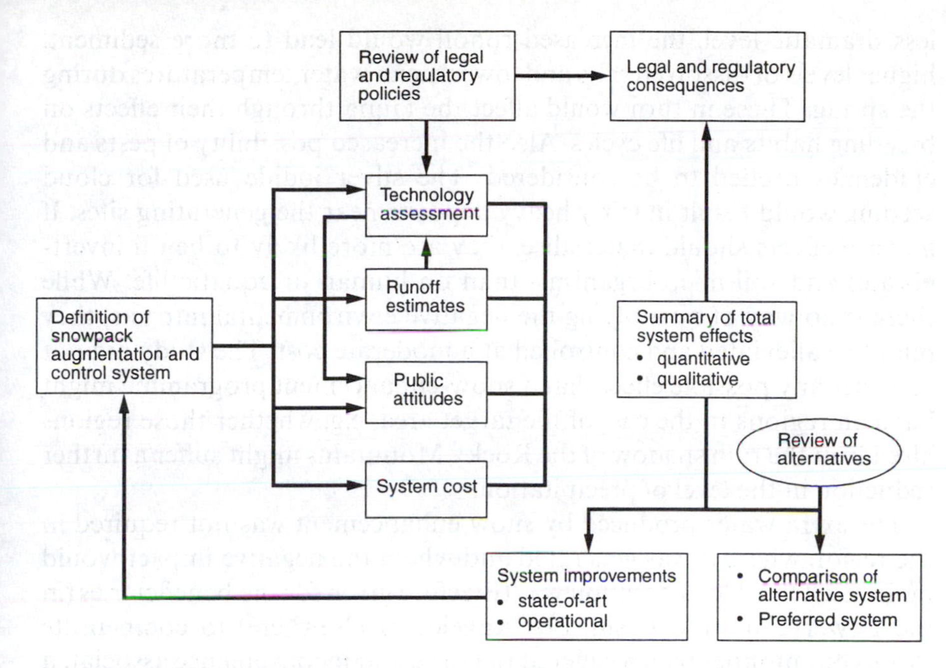

Figure 14.2 *Schematic approach to technology assessment of winter snowpack augmentation (climatic modification)*

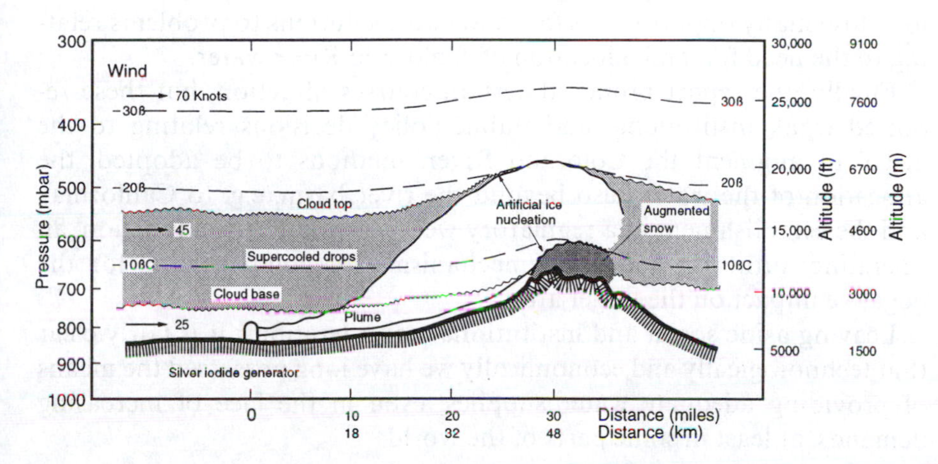

Figure 14.3 *Model showing the effects of cloud seeding on snowfall*

less dramatic level, the increased runoff would lead to more sediment, higher levels of dissolved salts and lower mean water temperatures during the spring. These in turn would affect the fauna through their effects on breeding habits and life cycles. Also the increased possibility of pests and epidemics needed to be considered. The silver iodide used for cloud seeding would result in fairly heavy deposits near the generating sites. If adverse effects should materialise, they are more likely to be on invertebrates and soil microorganisms than on human or aquatic life. While there is no way of quantifying the negative environmental impacts, they might be alleviated and controlled at a moderate cost. The study did not consider any possible effect that a snow-enhancement programme might have on regions to the east of the target area, e.g. whether those regions already in the rain shadow of the Rocky Mountains might suffer a further reduction in the level of precipitation.

The extra water produced by snow enhancement was not required in the region where it was generated and where the negative impact would make itself felt. Consideration was therefore given for the beneficiaries (in the Lower Colorado Basin, Los Angeles or elsewhere) to compensate those communities which suffered the risks and inconvenience associated with artificially increased snowfall. A survey of the population involved demonstrated that there was a general distrust of the federal agencies; also there was the belief that interference with natural weather patterns was immoral and smacked of witchcraft. While bearing in mind such public attitudes, the report suggested that Congressional inaction would foreclose many opportunities for innovative solutions to problems relating to the need for, and allocation of, Colorado River water.

Finally, the report proposed certain courses of action, but these required legal, institutional and public-policy decisions relating to the 'need' to augment the Colorado River, methods to be adopted, the allocation of the water (also beyond the river basin, e.g. to California), and the establishment of a regulatory weather modification function, an operating authority and some mechanisms for compensating for the negative impact on the target area.

Leaving aside social and institutional considerations, it is fairly clear that technologically and economically we have long possessed the means of providing adequate water supplies, even in the face of increasing demands, at least in some parts of the world.

Natural Water Quality

In the developed world, human preoccupation is often with the quality of potable water and the upgrading of the municipal water supply to the requirements of medicine and industry. Mention must, however, also be made of other important standards of quality, such as those that apply to agriculture, wild life, aquatic life, industry and recreation.

In the developed nations the high standard of potable water purity is taken for granted. The consequences of a lack of hygiene and even the present-day shortage of clean water in many parts of the world can hardly be exaggerated, and they will become more widespread. Even now, some 2.9 billion people, 66% of the population in the developing world, have no access to a toilet, not even a decent pit latrine. Overall, progress is being made in supplying potable water: in the five years from 1990, the proportion of people in the developing world with access to safe drinking water rose from 61 to 75%. Experts believe that, in the developing countries, it would be much more efficient to protect the sources of clean water than to attempt the cleanup of wastewater.

Urban slums present the worst sanitation nightmare, where death by water claims many lives, especially those of children. Estimates vary, but diarrhoea, schistosomiasis, trypanosomiasis and intestinal helminth infection account for 3.5 million lives annually. Where there is an insufficient supply of safe potable water, the population is exhorted to boil all water, but the monthly cost of fuel required might amount to a third of a family's monthly income, presenting them with difficult choices.

The most impressive progress in promoting community sanitation projects has been made by Socialist countries, notable Cuba and China. Thus, Cuba has achieved a life expectancy of 76 years, about the same as the USA. Shanghai has no India-style slums and much higher standards of hygiene. In fact, the life expectancy of a newborn child in Shanghai is longer than that of a baby born in New York City.

Education has played a major role in the achieving improvements. Many development economists believe that lives would be saved and health improved by offering credit to poor people to help them start businesses and generate some cash. This, in turn, would enable them to build covered wells and make some rudimentary sanitation improvements, since one of the most persistent reasons for poor hygiene is simply that there is almost no water to wash with in many parts of the world. There is no doubt that governments could save lives by supplying water to poor areas, not just clean water, but sufficient quantities, even if it means charging for the water to pay for the service.

Setting aside the gigantic problems associated with human pollution which, fortunately for many of us, hardly affect the Western world, the fight against industrial pollution of the environment has become a favourite cause of many special-interest groups. Society in the affluent countries has also become aware of the dangers accompanying economic growth. Many books have been written about the so-called 'gross national by-product'. This is not the place for a detailed discussion of that topic. However, any account of the turnover and utilisation of water would be incomplete without some comments on anthropogenic changes in water quality and the impact of changing health, agricultural and industrial practices.

There are essentially two kinds of chemical pollution of an aquatic environment, be it a river, ocean, or an underground deposit. Organochlorine pesticides are examples of one class of pollutants; these occur when technological advance produces a new material that is not known in nature. If such a substance is also chemically stable in the environment, or if it is only partly degraded to another, stable form, the situation is further complicated. Some such chemicals may be used because of their toxicity to certain species of animals or plants, in which case there are obvious dangers to other forms of life. This recognition led to the development of ever more refined and specific types of pest control.

Fortunately the application of toxic agrochemicals is usually carried out under the supervision of qualified staff, and their effects are now carefully monitored. The same is hardly true for industrial wastewaters which are still being discharged on an enormous scale. For every tonne of paper that leaves a paper mill, for example, 250 t of wastewater are produced, and although effluents are treated before discharge, many industrial processes give rise to highly toxic materials in trace concentrations, which are hard to remove completely. The incidence and levels of heavy metals in marine life bears witness to this. On the other hand, there can be little doubt that since the inception of municipal water treatment, the waterways and lakes have become very much cleaner than they were 150 years ago. On balance, there has been a decided improvement in the quality of our waterways.

Like the application of pesticides and herbicides, the treatment of industrial waste and its eventual discharge are also subject to some control, even if not to the same extent, but no regulation can prevent cases of inadvertent, large-scale pollution. There have been several incidents of massive oil pollution of oceans and beaches, usually caused by human error, the total cost of which in terms of environmental and economic resources is difficult to appreciate.

The effects of persistent industrial chemicals on aquatic life can be particularly serious, because of the low solubility of many organic compounds. Instead of dissolving and thus existing in a very dilute solution, they are concentrated at the surface of quiescent waters. The layers so formed are very thin, possibly only one molecule thick, but their presence drastically modifies the surface tension and the rate at which oxygen is taken up by the water. Both of these changes can have severe consequences on aquatic life.

The second type of pollution is caused by substances that serve as nutrients in the biosphere, i.e. they exist in nature and can be accommodated and metabolised by living organisms. Such substances, although essential for the maintenance of life, present problems when their concentrations in a given environment exceed certain levels. They can then disrupt biological equilibria, e.g. the interactions of populations of different plants and animals with each other. This is what occurred in Lake Erie during the 1950s and 1960s. The ecological disruption was due to an introduction of large quantities of nutrient materials for plants. While these are necessary for growth and health of the aquatic biological community, in high concentrations they cause cancer-like overgrowth, leading in extreme cases to a collapse of the ecosystem. Both inorganic and organic compounds can be involved in this way.

Nitrogen- and phosphorus-containing compounds are particularly important in this context. Just as the water in the hydrosphere is turned over continuously by the hydrologic cycle, so nitrogen from the atmosphere is 'fixed' by biochemical reactions to yield ammonia and more complicated organic derivatives of ammonia. Nitrogen is also oxidised to nitrate by complicated organic derivatives of ammonia. Simultaneously, nitrogen is oxidised to nitrate by various photochemical or bacterial processes. In turn nitrate in plants is reduced to lower oxidation states, and eventually to ammonia, by various enzyme-catalysed reduction steps. This is the most important part of the nitrogen cycle, because it forms the link between the organic and inorganic environments. The nitrogen cycle involves the atmosphere, living matter, decomposing bacteria, soil and water. It is easy to see that any gross perturbation, e.g. in the quantity of nitrate or waste products, is likely to lead to significant changes in the nitrogen balance and the growth patterns of certain organisms.

Similar considerations also apply to phosphorus and its turnover in the ecological environment. Its nutritional role is crucial to both plants and animals. An insufficiency of nitrogen or phosphorus can limit the growth of an individual or a community of plants, and their supply is regulated by the application of fertiliser to stimulate plant development. However, this very same process can be detrimental in an aqueous environment.

The nutrient requirements for plants can be summarised as follows: (i) light energy for photosynthesis, (ii) water and carbon dioxide as photosynthetic reagents, (iii) phosphate and nitrogen (usually as nitrate) and (iv) small amounts of many other minerals. Depending on the particular conditions, one or another of these will be the *limiting nutrient*, the concentration of which determines the overall growth rate. If this concentration is increased because of human activity or by some other change, then the plant population will increase. Aquatic plant overgrowth occurs because of such nutrient enrichment.

Three types of species are involved in the maintenance of the normal ecological equilibrium of a lake. The *producers* are the plants, principally algae, which take up carbon dioxide and produce carbohydrates. The *consumers* are the herbivores – and indirectly the carnivores – which feed on the plants, regenerating carbon dioxide through respiration. The *decomposers* interact with dead plant and animal tissues, and with wastes from living plants and animals, to regenerate carbon dioxide and inorganic components, including nitrate and phosphate, which are then available for recycling by the producers and consumers. In the process the decomposers (microorganisms) consume oxygen, as do the consumers. Energy enters this cycle from outside by means of photosynthesis, and so do dissolved inorganic nutrients.

Human interference – often called *development* – in the ecological cycle need not have serious results, but it will enhance the turnover rates and quantities of the various species involved. In extreme cases, however, the oxygen requirement of the plant decomposers may outstrip the oxygen-producing capacity of the plant producers. The oxygen supply then becomes too low for the consumers to survive in adequate numbers. Only anaerobic bacteria and algae survive. Plant decomposition slows down and the lake becomes more like a swamp than a body of water. This process, which is termed eutrophication, can be regarded as ecological obesity – an artificial ageing of the lake. Thus, in the last 70 years the discharge of nutrients into Lake Erie caused the ageing process to advance to a point that would not have occurred naturally for 20 000 years. In the years 1942 to 1967 the available nitrogen content (as ammonia or nitrate) increased by 50% and the soluble phosphorus content by a staggering 500%. Similar problems arise wherever lakes become 'industrialised', be it by manufacturing industry, agriculture, or domestic development, e.g. Lake Baikal and the Caspian Sea in Russia and Lake Tiberias in Israel.

Table 14.1 shows the annual levels of nitrogen and phosphorus enrichment of Lake Erie at the high point of the crisis, during the late 1960s. Detergents are the largest single source of phosphate, which is concentrated in the sewage from urban areas. The annual release of phosphates from detergents in the United States is estimated at upward of a million tonnes.

Animal wastes present essentially the same problems as human wastes, except that they are not likely to be treated as are domestic wastes. Thus the domestic animal population of the Unites States produces the waste

Table 14.1 *Phosphate and nitrate sources in Lake Erie (in 10^3 tonnes per year)*

Source	*Phosphate*	*Ammonia/nitrate*
Runoff from rural urban lands	4.3	68
Detergents	11.6	–
Human and livestock waste	5.0	31.3
Added in rainfall	–	12
Fixation by soil organisms	–	25.7
Industrial discharge	1.0	?

equivalent to 2 billion humans. Spread over a large area this does not create many problems, but the recent tendency has been to concentrate livestock and poultry in densely populated feed lots.

The problems associated with the treatment and disposal of human waste are of major dimensions. The total domestic effluent in the United States amounts to ca. 1.5 million ha m which contains only 0.5% solid waste. Thus a community of 25 000 will produce approximately 5 million l of raw sewage a day. This is equivalent to the effluent produced daily by a medium sized brewery. While primary filtering removes suspended solids, a secondary microbiological treatment is frequently applied to remove soluble organic materials. At this stage the effluents contain high levels of carbon dioxide and low levels of oxygen. In addition, none of the inorganic products have been removed. Yet this is the point at which most sewage treatment operations terminate, although technically the means exist for substantial reductions of nitrate and phosphate levels in treated water.

With a still growing agrochemical industry, the total annual synthetic production of ammonia now rivals the activity of leguminous plants in converting atmospheric nitrogen into a form that can be assimilated by living organisms. Although serious nutrient water pollution could arise if the fertilisers were washed out of the soil by rainfall, it has been shown that phosphates are largely rendered insoluble by reacting with soil particles, so that runoff does not present major problems. The situation with nitrates is rather different because of their high solubility. Although it has not yet been established how serious this is in connection with eutrophication, very high nitrate levels in potable water supplies can give rise to other problems, and have been known to do so. For example, the pathological condition known as methaemoglobinaemia results indirectly from exceptionally high nitrate levels in drinking water. The effects, which are akin to those of carbon monoxide blood poisoning, result from the bacterial reduction of nitrate to nitrite and the ensuing effects of nitrite on haemoglobin. The physiological conditions under which the nitrate-reducing bacteria can exist only occur in infants, who are therefore at risk from high (approximately 100 mg^{-1}) nitrate levels in water. Unfortunate combinations of circumstances involving certain irrigation practices and usage of fertilisers have in the past led to some known cases of the disease in different parts of the world.

The quantity of nitrogen released into lakes is much larger than that of phosphorus. The question remains as to which of the two elements might be the limiting nutrient. Certainly much more nitrogen is required for plant growth than phosphorus, but there are still some remaining doubts as to the specific cause of eutrophication in Lake Erie or elsewhere. It is clear, on the other hand, that once the nutrient levels are reduced, a lake or stream can be gradually restored to its former condition.

Finally, there is one other process which can be considered under the general heading of pollution: the salination of underground water. Problems arising from the widespread nature of saline soils can be further compounded by the geographical distribution of populations and by agricultural practices that have largely succeeded in increasing salination in arid and semi-arid lands. Although only Europe and Australia have been mapped in detail, it is estimated that at least 4 million km^2 of the earth's land surface is affected by salinity, excluding the areas of the major deserts.

Probably the effects of secondary salination are even more serious, since they represent losses of once-productive agricultural land. In areas such as the Punjab, 25% of the 51 000 km^2 of agricultural land had already been seriously affected by salinity in 1960, while some 10% had been lost to agriculture. Land was going out of use at an annual rate of 400 km^2. Such losses are due primarily to irrigation, the scale of which can be judged from its current consumption of 80% of the world's total water supplies during the farming season. Water used for irrigation is often of a poor quality, so evapotranspiration leads to the concentration in the soil of salts added to the land during irrigation. In the estimation of some experts the spread of salinity in irrigated lands is so severe that they question the long-term viability of irrigation schemes unless agricultural practices are considerably improved. Since fast improvements are improbable in the developing countries, the biological solution to the problem, already mentioned earlier, inducement of salt tolerance in plants, becomes of greater urgency.

Water Purity and Purification

The term 'quality', as applied to water, must take account of the end use. Thus, the quality of potable water will be governed to a large extent by toxicological and microbiological criteria, which have been laid down by national or international regulatory authorities. The methods adopted depend largely on the quality of the local ground and/or surface water resources.

In East Anglia, one of the more arid regions of the British Isles, drinking water is obtained in equal parts from surface and ground reservoirs. The gradual establishment of European directives, coupled with the privatisation of the water supply companies, fuelled investment aimed at upgrading the quality of potable water. The primary aim at the time was the reduction of lead from 50 to 10 $\mu g\, l^{-1}$. The method adopted was based on the precipitation of lead with orthophosphate, although subsequently, in another part of the cycle (sewage treatment) the lead has to be removed.

Another European directive deals with nitrate levels. They should be kept low mainly by reducing the nitrogen input into potable water sources. The directive lays down a level not exceeding 50 $mg\, l^{-1}$. Finally, the only possible microbiological hazard in the UK, cryptospiridium, is removed from drinking water by filtration. The UK now conforms with all the European directives. In East Anglia this is achieved by a combination of ozone treatment, charcoal adsorption, filtration and the addition of phosphate and coagulants. In the Netherlands, 'natural' filtration is employed, by pumping water through sand dunes.

No method is completely foolproof, as witnessed by the highly publicised disaster in Cornwall, where, on 6 July 1988, high levels of aluminium found their way into the drinking water. The possible implications with Alzheimer's Disease featured prominently in the news media at the time. Altogether the European Community influence has been beneficial and has led to major improvements in water purification technology, although serious problems still exist with the treatment and discharge of sewage at many European beaches.

The treatment of ultrapure water, as required by the semiconductor industry, must be judged against a much more stringent set of criteria. Water quality can be expressed in units, such as electrical conductivity or resistivity ($\mu S\, cm^{-1}$), total dissolved solids ($mg\, l^{-1}$), pH, etc. In 1894 Kohlrausch prepared what later came to be known as 'conductivity water' (0.055 $\mu S\, cm^{-1}$ at 25 °C), by a series of 26 distillations in quartz. The results were to be an acceptance of conductance as a valid measure of purity, coupled with an erroneous conclusion that conductance can be accepted as evidence for the absence of *all* dissolved or suspended impurities. Water with a low conductance, prepared by ion exchange, can nevertheless have a depressed surface tension, indicating the presence of appreciable concentrations of absorbed surface-active substances of a nonionic nature. Depending on the origin of the feed water, such contaminants can consist of humic and/or folic acid decomposition products. The water, although possessing the features of 'conductivity water', will then also decolourise potassium permanganate!

A discussion of quality may take as a useful starting point the classification of natural water. The most common example is the classification into 'soft' and 'hard' waters. Hardness is usually expressed in terms of calcium and magnesium ion concentrations which cover a range of 60–180 ppm. The nature of the anion determines whether hardness can be removed by boiling (temporary hardness), e.g. carbonates and bicarbonates, or whether the hardness is permanent, e.g. sulfates and chlorides.

By contrast to natural waters, no quantifiable definition exists for purified water, even for potable water. Indeed, water from many natural wells, containing large amounts of dissolved minerals and gases, is regarded as wholesome and even as health improving when sold in bottles. Such 'natural' water would be completely unacceptable for science, medicine and industry, where it could only serve as feed water for further purification. The actual selection of purification methods must take into account the feed water quality and, in particular, the nature and concentrations of dissolved and/or suspended substances.

Common industrial purification processes include filtration, distillation, ion exchange, reverse osmosis (desalination), electrodialysis and exposure to UV radiation. In order to remove colloidal impurities, various pretreatments designed to promote particle aggregation can be employed; they include flocculation by polyvalent ions, freeze/thaw cycles, pH adjustment, gravity settling and use of ultrasonic radiation. Such treatments are commonly applied in wastewater processing, where the emphasis is not so much on the purification of bulk water as on the dewatering of sewage sludge.

Since water is an almost universal solvent, the storage of purified water may also present problems in some industries. In fact, storage is best avoided, unless the rate of production cannot be exactly matched to the demand. Where storage is unavoidable, highest priority must be given to the exclusion of atmospheric contamination, because airborne particles will quickly reduce water quality. Other factors that require attention include the continuous motion of water in the storage tank, tank design, construction and ancillaries, e.g. pumps and piping, and cleaning procedures.

Pure Water in Medicine and Industry

Pharmaceutical manufacturers, hospitals and other medical establishments require purified water of different grades. The most stringent

Table 14.2 *Empirical criteria for ultrapure water in the semiconductor industries*

Parameter	*Quantity and unit*
Electrical conductivity	0.055 $\mu S\ cm^{-1}$
Particle count	100 ml^{-1}
Particle size	0.5 μm
Organic content	1 ppm
Bio count	1 ml^{-1}

Table 14.3 *Quality of water sample from 'pure' feedstock and after purification for semiconductor production*

Parameter	*Input water*	*Final product*
pH	7.7	5.5
Odour	None	None
Turbidity	0.0	0.0
Colour, Pt-Co units	0.0	0.0
Ammonia as N	0.10	0.03
Nitrite as N	0.001	0.000
Nitrate as N	0.23	0.00
Alkalinity as $CaCO_3$	80.0	0.47
Hardness as $CaCO_3$	204.0	0.00
Hardness, g/gal	11.9	0.00
Calcium as $CaCO_3$	160.0	0.00
Magnesium as $CaCO_3$	44.0	0.00
Calcium as Ca	64.0	0.00
Magnesium as Mg	10.65	0.00
Total solids	340.0	0.8
Fixed solids	256.0	0.2
Volatile solids	86.0	0.6
Iron	0.044	0.009
Manganese	0.005	0.000
Aluminium	0.06	0.00
Silica as SiO_2	12.42	0.043
Chloride	38.0	0.45
Fluoride	0.075	0.000
Sulfate	120.0	0.0
Sodium	20.55	0.003
Potassium	1.88	<0.006
Oxidisable matter	None	None

Concentration of dissolved substances are given in ppm.

criteria apply to the grade 'water for injection'. Sir Christopher Wren, assisted by Robert Boyle, appears to have been the first person, in 1656, to attempt intravenous injection into animals, by means of a hollow quill. The hypodermic injection of analgesic fluids into humans dates from 1855 and parenteral drug delivery is now a common clinical technique. Water is the principal vehicle for most preparations, and the purity requirements for aqueous solutions include compatibility with blood; effective solubility of solid components; physical, chemical and microbiological stability; an absence of toxic principles; and that they should be non-sensitising, non-irritating and unaffected by pH changes. Among factors inducing toxicity, a group of substances which became known as pyrogens are the most important and also the most difficult to remove from otherwise pure water.

Gram-negative bacteria appear to be the most potent producers of pyrogens, which can have profound effects in nanogram quantities. Such endotoxic effects manifest themselves as skin irritation, followed by headache, muscle pains, nausea and even more severe symptoms. For obvious reasons, the permitted methods for the preparation of 'water for injection' are now strictly regulated; in the main they are limited to distillation or reverse osmosis, followed by autoclaving.

The electronics industry requires ultrapure water for the cleaning of active crystal surfaces, by the removal of alkali and heavy metal ions. Applicable purity criteria are summarised in Table 14.2. A typical purification train might consist of a passage over charcoal beds, followed by electrodialysis to remove dissolved solids and pumping through mixed bed polishing units to remove residual impurities. Finally, the purified water is passed through a bank of 0.2 μm filters. The degree of purification of 'pure' feed water, subjected to the above treatments is shown in Table 14.3. Alternative processes, employing reverse osmosis, achieve similar degrees of purity. The choice of a suitable process is usually based on the required throughput and economic criteria.

Chapter 15

Economics and Politics

Economics of Water Consumption

Much is still being written about water demand and supply, but here again national or global averages present a distorted picture, because regional fluctuations from the mean can be quite large. It is interesting, though, that in Europe domestic consumption stands at about 230 l per person per day, which is what it was in the days of Imperial Rome. Presumably it dropped during the Middle Ages and rose again during the past two or three centuries. What has changed in 2000 years is the total population on earth and the growth of a water-consuming industry. In fact, the water usage of a nation, which is a gauge of its state of industrial development, tends to rise by 2–3% annually. This raises the question of a potential water shortage in the future. In the United States the potential supply is roughly equal to the average runoff (4.5 billion m^3 per day). The present demand amounts to about 75% of the potential supply. Of the demand, less than 30% is consumptive, the remainder being returned to watercourses to become available for further use. Given a constant rate of increased usage, demand will outstrip supply during the twenty-first century. On the other hand, the nation's water supply may be considered unlimited for several reasons:

(1) A large part of what is now classified as consumptive use is 'consumptive' only because facilities do not exist for reclaiming the water.
(2) Evapotranspiration is responsible for a large part of consumptive use. The efficient control of water-loving plants, which grow along rivers and reservoirs, has already resulted in a reduced water loss. Further reduction can be effected by controlling evapotranspiration in agricultural irrigation, making use of currently available

knowledge of the adaptation of crops to moisture constraints. Suppression of evaporation from reservoirs by reducing water-surface exposure also holds promise.
(3) Improved irrigation can reduce seepage conveyance losses.
(4) Weather modification may add considerably to usable water supplies at some future date; this aspect is discussed further below.
(5) The conversion of seawater to fresh water may be uneconomical now, but improved processes coupled with necessity in certain areas will make this a practicable option, as it already is in the Gulf States.

It is sometimes stated that the United States or some other developed countries are suffering, or will suffer, from severe water shortages. To counter this by stating the fact that the supply of water is virtually unlimited is only half-correct.

Additional supplies of water can be obtained only through the mobilisation of increasingly large amounts of our resources. Desalination of seawater, now technically feasible, could take care of human requirements for ever. However, the cost of water produced from the sea is still very much higher than that of water obtained from other sources. In contrast, the evaporation of fresh water can be prevented by spreading one-molecule-thick layers of substances, such as cetyl alcohol, on the water surface, at considerably lower cost than distillation. Even this procedure costs very much more than freshwater production.

What applies on a continental or national basis is equally true for smaller regions or communities. Thus, individual cities or regions may well experience periods of rapidly increasing demand, which could result from industrial or domestic developments, but there can be no 'shortage' of water. The exploitation of the available resources will become more costly as future demands increase, but we can obtain all the additional water desired, provided that we are willing and able to pay the high cost.

Water as an economic good or resource therefore exhibits the essential qualities of all other economic goods or resources. Water exists abundantly in nature. Nevertheless, it can be obtained for human use only through the application of varying amounts of scarce resources. Like the production of other economic goods, water production must eventually obey the law of diminishing returns. For any given location and state of water-production technology, an increased water output is ultimately associated with increasing marginal and average costs. Society can develop additional water supplies, if it so desires, but only at a higher unit cost.

Although to emphasise that water is like other economic goods may seem obvious, historically this has apparently never been understood.

Even economists themselves have seldom recognised the close relationship between water and other goods. There is evidence that today's professionals whose responsibility is the provision or allocation of water supplies also feel that, somehow, water is 'different'.

The misunderstanding probably springs from two truths:

(1) Water is abundant
(2) Humanity cannot survive without water.

These observations probably prompted Adam Smith's distinction between 'value in use' and 'value in exchange', expressed in his 'diamond–water paradox'. He supposed diamonds to have a market value, but no value in use, and water to have a high value in use, but no value in the market place. That people cannot live without water has been largely responsible for laws that have placed water apart from other economic goods, and which have little bearing on the conditions for optimum allocation of resources. Economic efficiency requires that:

(1) the existing water supplies be allocated in a way that will maximise their contribution to social welfare;
(2) a proposed water project be foregone unless it is warranted by the demands of society;
(3) any project that is undertaken be the least costly of the available alternatives.

An examination of various types of national water legislation demonstrates that some existing laws, and also the common practice of selling water on a flat-rate basis, are hardly in harmony with these requirements. Until the institutional setting is altered, the misallocation of water – and particularly ground water – will continue.

Human Attitudes and Politics

Part of the rainfall from year to year is systematic and should therefore be predictable, but much work remains to be done before reliable predictors can be established. Even if they were to be available, it is doubtful whether they could be exploited widely. Subsistence farmers could hardly respond to forecasts of the character of a season; their agricultural methods and attitudes are not sufficiently flexible. This was summed up succinctly in the *Ecologist*: 'Scientists and development planners work out elaborate schemes for rural regeneration, but peasants and goats seldom seem to find it in their own interest to pay attention to the profile of the computer printout.' The implied criticism is valid, but in a choice between short-term expediency and longer-term benefit, governments, even if not the individual farmer, should be able to respond to predictions of the character of a season. If a particularly dry season is forecast, food can be stored and cattle can be limited by controlling access to wells. Migration of people away from the areas most at risk can be encouraged and assisted.

Probably most importantly, governments could develop some independence from climatic fluctuations by coordinating economic and other exchanges with other climatic areas. Since rainfall patterns change quickly with distance in the tropics, seasonally arid and humid regions are often found close together. Before West Africa was divided up between the colonial powers, the wet and dry regions had traditionally been linked by reciprocal trading systems. These arrangements were further developed during the colonial period. The more recent national structure of West Africa has resulted in political barriers across these natural north–south links, thereby isolating the northern countries with their unreliable rainfall patterns. Although politically independent, these countries cannot attain agricultural self-sufficiency at any but the poorest level; nor can they cope independently with the consequences of their erratic climatic conditions.

The most important differences in the exploitation of America's or Australia's arid and semi-arid lands in comparison with developing countries are thus seen to be social, economic and political. The Australian grazier runs a highly capitalised and highly mechanised business, well integrated with areas of higher rainfall through a communication, distribution, and market system. Success and failure are calculated in terms of profits. This is totally different from the subsistence industries of many other arid areas where one year's drought may mean starvation and death. Technically, there are few problems in adapting successful management practices to Third World countries because the physical limitations set by the distribution of water resources are very similar. Furthermore, the necessary data and experience exist. The major barriers to their implementation are of an economic, social, and, above all, political nature. The more urgent problems associated with the achievement of a balanced system of water management will therefore not be overcome by interdisciplinary research alone, although this is important. Rather, the information generated by international research activities must lead to local economic and social changes, through enlightened community behaviour and courageous political leadership.

Commercial Exploitation of Water Shortages

The economic benefits that might accrue from exploiting the world's water shortages have also not gone unnoticed. *The Independent on Sunday* newspaper of 3 October 1999 carried a page one banner headline: 'Monsanto plan to cash in on world water crisis'. According to a leaked confidential report, the chemicals company produced a 'water business plan', subsequently dropped, to make billions of dollars and, in the process, 'to help solve some of the world's major environmental issues and to improve quality of life in the process . . . doing the right thing while we are doing business'. The report is quoted as stating that '. . . population growth and economic development will apply increasing pressure on natural resource markets. Those pressures . . . will create vast economic opportunity.' *The Independent on Sunday* further quotes a comment by Dr Vandana Shiva, the director of the Indian Research Foundation for Science, Technology and Ecology: 'Monsanto is seeking a new business opportunity because of the emerging water crisis . . . Privatisation and commodification of water are a threat to the right to life.'

Although the company's plan for the exploitation of the world water situation has been shelved, it can only be a matter of time before a fierce competition will develop to derive profit from the growing water shortage. A minor forerunner of such international competition already took place some twenty years ago, when the Gulf States first decided to install large-scale desalination equipment.

Future Outlook

The continuous self-renewal of the world's freshwater supplies through the hydrologic cycle takes place at such a rate that there is no danger of any future global water shortage. However, because of spatial and temporal fluctuations in the levels of precipitation, local shortages will persist until humanity learns to develop some measure of climatic control. This is not likely to come about until a better understanding can be achieved of the complex relationships between meteorology, geology and agriculture. In the meantime, enlightened water-management procedures can do much to enhance the utilisation of semi-arid lands.

At the level of the individual organism, or indeed the single cell, water relationships in plants and animals are only partly explored. There exist many pathological conditions associated with disturbances in the circulation, distribution and functioning of body fluids.

Although most attention is usually paid to the improvement of potable water, it must always be borne in mind that for the balanced maintenance of life other quality standards are of equally great importance. This lesson has apparently been taken to heart in the more affluent countries of the world. Water is a primary and indispensable natural resource; the preservation of adequate supplies of acceptable standards of purity is not easily compatible with economic growth and all it entails in changing living patterns. Thus, even with no shortage of water in the hydrosphere, an increasingly heavy price will have to be paid to ensure continued supplies of this precious liquid which we have taken for granted for too long.

Over the past 25 years the world has experienced the impact of oil price fluctuations on social and economic developments. More often than not, such fluctuations were brought about by wars. There can be little doubt that in the medium-term future, the availability and cost of potable water will have an even greater impact on social stability, especially in the less-developed, semi-arid regions. Even now, 7% of the human race does not have enough water for a healthy survival, and this is likely to rise to a staggering 70% by the year 2050 unless urgent remedial measures are soon put in place.

UNESCO has agreed to establish a World Hydrology Initiative, one of its objectives being to curb the spread of 'water wars'. Political battles are already spreading across Africa, the Middle East and Central Asia. The world's annual population increase is the equivalent of a new India being added every ten years. The irony is that the world is awash with water, of which 97% is unfortunately seawater. Most of the remainder is trapped underground or locked up in the Antarctic ice cap. Over the years the towing of icebergs to Southern Australia may become reality.

The greatest need for water is for farming. About 75% of the world's water supply is used for growing food, but in order to utilise more of the annual precipitation, significant advances in irrigation technology need to be made. Most such advances can hardly be described as high-tech. They include leaving the ground covered with leaves and mulches, the planting of mixes of crops and bushes and making improvements in water storage.

Recently the United Nations adopted a convention on international waters, with the aim of providing a framework for the sharing of 'international' water. Only 11 countries have signed up, and the convention is likely to go by default. Water remains a political issue but is central to any attempt to end extremes of poverty. A major problem is that the need for water is most acute in countries that are threatened with having their supplies controlled. Thus, the Nile cannot even now supply Egypt with all its water needs, but upstream Ethiopia, Somalia and Sudan have begun the building of dams because they, too, require more water. Egypt, whose population will double in 35 years, has made it clear that the diversion of more Nile water would be regarded as an act of war. A similar potential flashpoint is provided near the border of Turkey, Iraq and Syria, where Turkey plans to divert water from the Tigris for agricultural and domestic development. The President of Turkey has put it bluntly: 'Neither Syria nor Iraq can lay claim to Turkey's rivers any more than Turkey can claim their oil.' Other major rivers whose exploitation might cause conflict include the Zambesi and the Mekong

On the bright side, Nepal, Bangladesh and India recently signed an agreement over sharing the water of the Ganges. One must hope that future water wars can be avoided by generating the means of decreasing the proportion of plentiful precipitation that is now irretrievably lost in the oceans. This sufficiency of supply does, however, need to be coupled with adequate provisions for maintaining acceptable standards of quality and public hygiene, especially in densely populated regions.

Chapter 16

Summary and Prognosis

Nobel Laureate A. Szent-György called water 'the matrix of life'. Among chemical substances it is unique in many respects, despite its apparent molecular simplicity. After sixty years of intensive study we have still not succeeded in fitting its physical properties into a general classification of liquids. Despite symposia and many publications, purporting to deal with 'Water in Biological Systems', we really have next to no understanding what makes the only inorganic liquid, occurring naturally on our planet, the absolute prerequisite for organic Life, as we know it.

Since the initial publication of the RSC Paperback *Water*, progress has certainly been made in some areas, but some of the more recent discoveries have also added to the mysteries that remain. It is beyond doubt that the development of neutron scattering techniques has led to a much better insight into the nature of so-called hydration interactions and has also been useful for testing hydration models, arrived at by computer simulation or *ab initio* theoretical approaches. Similarly, modern NMR techniques have made possible the detailed study of the role played by water in the promotion and maintenance of complex biological structures. There can no longer be any excuse for omitting the contribution of water from energy calculations on biopolymers.

Little progress has been made in the development of an improved quantitative description for the liquid substance on a molecular basis. One reason is surely the necessity for treating a many-body problem in terms of the sum of two-body interactions. There are as yet no satisfactory analytical methods for the description of three-body interactions. Also there is the complexity of the water molecule (three nuclei and 10 electrons) which rules out rigorous *ab initio* aproaches. We have at our disposal a plethora of potential functions for the water dimer, none of which is, however, completely satisfactory.

Computer simulation is presently the most popular method for the study of complex aqueous systems. In the hands of the sceptical and modest expert, simulation is a powerful tool. In other cases extreme caution must be exercised before accepting results from such calculations, especially where they cannot be tested experimentally.

The intermolecular nature of water first became a subject of study during the early years of the last century and reached its zenith during the middle of that century. It is a common fallacy that new interpretations reported in the immediate past are necessarily more correct than those published, say 50 years ago. Nothing could be further from the truth. Like every other faith, 'water structure' continues to have its share of misinterpretations and false prophecies. In an era of rapid publication, it is impossible to judge what will persist and what will be regarded as a dead end in fifty years' time.

On the positive side, the understanding of ion hydration has been vastly improved. Neutron scattering has enabled us to develop a detailed description of primary ionic hydration shells. The range (thickness) of hydration shells remains to be established. This will require improved methods for analysing more distant peaks in atom–atom distribution functions. The origin of the lyotropic ionic series remains a major mystery. Although, since its discovery in 1888, it has been rediscovered at least once in every decade, we still await a satisfactory explanation. Bearing in mind that the ionic series applies not only to stabilities and solubilities of complex molecules, such as proteins and surfactants, but also governs the solubilities of noble gases, any credible explanation cannot rest on specific properties of the complex substrate molecules. Hofmeister, in his seminal publication of 1888 speculated that the series was '. . . probably related in some way to the affinities of different ions for water.' We have not really advanced much beyond that insight.

Earlier pioneering studies on undercooled water brought to light an apparent singularity at −45 °C, characterised by the divergence of most physical properties. This phenomenon provided further impetus to studies of undercooled water, especially when coupled to the realisation that liquid water could be quench-cooled into an amorphous solid. Two amorphous phases, differing in their densities, have been identified and studied in detail. The possibility of two coexisting liquid water phases continues to attract great research interest. The implications of liquid-state polymorphism are exciting, particularly to theoretical physicists and computer simulators. It must be remembered, though, that the existence of different liquid phases is not peculiar to water. A similar liquid–liquid transition has been reported for phosphorus and suggested for carbon. The indications are therefore that a second critical point should exist, in the neighbourhood of −50 °C. It lies in a temperature region, approximately -40 to $-140\,^{\circ}C$, that is still inaccessible to experimental study, because of the nucleation and rapid growth of ice. Therefore, although probably not of any immediate practical significance, the research future of unstable water at low temperatures and/or high pressures is assured. Any future discoveries may shed light on the phase behaviour of 'real' materials that display a tetrahedral structure, similar to that of water, notably SiO_2 and GeO_2. The unique feature of water is, of course, that such structures are not produced by covalent bonds, but by weak hydrogen-bonds with short lifetimes and high exchange rates.

It is somewhat surprising that earlier intensive research into supercritical water has not continued at the same pace. At one time it was believed that supercritical water, like supercritical carbon dioxide, might have a bright future in chemical processing. Its remarkable solvent properties and the absence of environmental pollution hazards should make supercritical water a favoured solvent for chemical processing. Engineering feasibility studies are on record, for instance, for the depolymerisation of cellulosic waste to glucose, a process that can be carried out rapidly, without charring. Probably the high temperature and pressure required for such processes outweigh any perceived advantages.

Related to the subject of 'unstable' water, amorphous solid phases containing low amounts of water continue to generate research interest. In particular, the dynamics of vitreous PHCs and of water within such phases appear to be of importance in pharmaceutical and food process technology, where they govern long-term product stability. Another topic, also related to amorphous aqueous phases, is the mechanism of desiccation stress resistance in ectotherms.

The role of water in the promotion and stabilisation of functional biological molecules also deserves more intensive study. It is questionable whether any more useful information can still be derived from crystal structures. On the question of biological recognition phenomena, such as occur between living cells, it is likely that hydrated, rather than anhydrous, sugar residues are involved. With the subject of PHC hydration still in its infancy, it is unlikely that rapid progress will be made in elucidating the molecular mechanism of biological recognition.

At the level of the individual living cell, or indeed for whole, land-based organisms, the water relationships that maintain biological viability are still largely unexplored. It is thus not surprising that there exist many pathological conditions associated with disturbances in the circulation, distribution and functioning of body fluids, or the unpredictable crystallisation of mineral 'stones' in organs such as the gall bladder, kidney or prostate. Also unexplored are the mechanisms whereby water participates in chemical and biochemical reactions. So often, deceptively simple equations disguise complex and intricate reaction mechanisms.

Water is a primary and indispensable natural resource, and globally there is no shortage of it in a hydrosphere that also provides for the continuous self-renewal of fresh water supplies through the evaporation/precipitation cycle. The regional use, availability, distribution and management of water resources are subject to complex relationships, many of them still unexplored, between geology, meteorology, glaciology, agriculture and economics, several of them subject to social and political overtones.

To place water in its rightful context, one can do no better than to quote James Joyce:

> *'Its universality . . . its vastness in the ocean . . . the restlessness of its waves . . . its hydrostatic quiescence in calm . . . its sterility in the circumpolar ice-cap . . . its preponderance in 3:1 over the dry land of the globe . . . its slow erosions of peninsulas . . . the simplicity of its composition . . . its metamorphoses as vapour, mist, cloud, rain, sleet, snow, hail . . . its submarine fauna and flora, numerically, if not literally, the inhabitants of the globe . . . its ubiquity as constituting 90% of the human body . . .'*

Suggestions for Further Reading

Many of the aspects of the physics and chemistry of water and of aqueous solutions were first treated in detail in a seven volume series entitled *Water – A Comprehensive Treatise*, published by Plenum Corporation, New York, completed between 1972 and 1982. Although many years later they may be primarily of historical interest, they are referred to below in roman numerals (Volume) and parentheses (Chapter). Some of those topics are also described in this book; especially where they have been further developed over the years. Other aspects of water have only recently been discovered or treated in any detail. Full references are given for these, but the bibliography is not, and was never intended to be exhaustive; emphasis is placed on significant reviews and books. It will only provide signposts to point the interested reader in the right direction for further study.

Chapter 1
Water, I(1).
B. Schröder, *Wasser*, Suhrkamp, Frankfurt, 1970.
F. Franks, 'The Hydrologic Cycle', in *Handbook of Water Purification*, ed. W. Lorch, McGraw-Hill, London, 1981.
M. Gross, *Life on the Edge*, Plenum Press, New York, 1998.
R.M. Kelly, J.A. Baross and M.W.W. Adams, 'Life in Boiling Water', *Chem. Britain*, July, 555, 1994.
L. Wallace, P. Bernath, W. Livingston, K. Hinkle, J. Busler, B. Guo and K. Zhang, 'Water on the Sun', *Science*, **268**, 1155, 1995.

Chapter 2
Water, I(1, 2, 3).
J.L. Finney, J.E. Quinn and J.O. Baum, 'The Water Dimer Potential Surface', *Water Sci. Rev.* **1**, 93, 1985.

Chapter 3
Water, I(1, 5, 10, 13, 14).

Chapter 4
Water, II(3).
P.V. Hobbs, *Ice Physics*, Oxford University Press, Oxford, 1975.
W.F. Kuhs and M.S. Lehmann, 'The Structure of Ice-Ih', *Water Sci. Rev.*, **2**, 1, 1986.
H. Suga, 'A Facet of Recent Ice Sciences'. *Thermochim. Acta*, **300**, 117, 1997.
J. Evans, 'The Molecules that Fell to Earth, *Chem. Britain*, **25**, 28, 1999.

Chapter 5
Water, I(14); VI(6).
J.C. Dore, 'Structural Studies of Water by Neutron Diffraction', *Water Sci. Rev.*, **1**, 3, 1985.
J.L. Finney and A.K. Soper, 'Solvent Structure and Perturbations in Solutions of Chemical and Biological Importance'. *Chem. Soc. Rev.*, **23**, 1, 1994.
F.H. Stillinger and T.A. Weber, 'Inherent Structure in Water', *J. Phys. Chem.*, **87**, 2833, 1983.

Chapter 6
Water, II(1, 5); IV(1); VI(5).
R.D. Broadbent and G.W. Neilson, 'The Interatomic Structure of Argon in Water', J. Chem. Phys., **100**, 7543, 1994.
A. Geiger, A. Rahman and F.H. Stillinger, 'Molecular Dynamics Study of the Hydration of Lennard–Jones Solutes', *J. Chem. Phys.*, **70**, 263, 1979.
D.T. Bowron, A. Filipponi, M.A. Roberts and J.L. Finney, 'Hydrophobic Hydration and the Formation of Clathrate Hydrates', *Phys. Rev. Lett.*, **81**, 4164, 1998.

Chapter 7
Water, III(1, 4, 6, 8); VI(1).
F. Hofmeister, *Arch. Exptl. Pathol. Pharmakol.*, **24**, 247, 1888.
C.I. Ratcliffe and D.E. Irish, *Water Sci. Rev.*, **2**, 149, 1986; **3**, 1, 1987.

Chapter 8
F. Franks and J.E. Desnoyers, 'Alcohol–Water Mixtures Revisited', *Water Sci. Rev.*, **1**, 171, 1985.
T.H Lilley, 'The Solvation of Amino Acids and Small Peptides', *Water Sci. Rev.*, **5**, 137, 1990.
F. Franks and J.R. Grigera, 'Solution Properties of Low Molecular Weight Polyhydroxy Compounds', *Sci. Rev.*, **5**, 187, 1990.
F. Franks, 'Protein Hydration', in *Protein Biotechnology*, ed. F. Franks, Humana Press, Totowa, NJ, 1993.

F. Franks, 'Protein Destabilization at Low Temperatures', *Adv. Protein Chem.*, **46**. 105, 1995.

Chapter 9
Water, VI(4).
W. Blokzijl, J.B.F.N. Engberts and M.J. Blandamer, 'Qantitative Analysis of Medium Effects on Organic Reactions in Mixed Aqueous Media', *Trends Org. Chem.*, **3**, 295, 1992.

Chapter 10
Water, IV(4, 5, 6).
J.H. Crowe and L.M. Crowe, 'Lyotropic Effects of Water on Phospholipids', *Water Sci. Rev.*, **5**, 1, 1990.
E. Westhof and D.L. Beveridge, 'Hydration of Nucleic Acids', *Water Sci. Rev.*, **5**, 24, 1990.
J.L. Finney, 'Hydration Processes in Biological and Macromolecular Systems', *Faraday Disc*, **103**, 1, 1996.

Chapter 11
M.W. Denny, *Air and water – the Biology and Physics of Life's Media*, Princeton University Press, Princeton, NJ, 1993.
A.D. Tomos, 'Cellular Water Relations of Plants', *Water Sci. Rev.*, **3**, 186, 1988.
F. Franks and S.F. Mathias, eds., *Water Biophysics*, John Wiley & Sons, Chichester, 1982.
R.M. Laws and F. Franks, convenors, Life at Low Temperatures, *Phil. Trans. Roy. Soc. B*, **226**, 515–692, 1990.

Chapter 12
Water, VII(1, 2, 3).
P.G. Debenedetti, *Metastable Liquids*, Princeton University Press, Princeton, NJ, 1996.
H. Tanaka, 'Phase Behaviour of Supercooled Water – Reconciling a Critical Point of Amorphous Ices with Spinodal Instability', *J. Chem. Phys.*, **105**, 5099, 1996.

Chapter 13
Water, VII(3).
J.M.V. Blanshard and P.J. Lillford, eds., *The Glassy State in Foods*, Nottingham University Press, Nottingham, 1993.
F. Franks, *Biophysics and Biochemistry at Low Temperatures*, Cambridge University Press, Cambridge, 1985.
H. Levine and L. Slade, 'Water as Plasticizer: Physico-chemical Aspects of Low-moisture Polymeric Systems', *Water Sci. Rev.*, **3**, 79, 1988.

E. Yu. Shalaev, F. Franks and P. Echlin, 'Crystalline and Amorphous Phases in the Ternary System Water–Sucrose–Sodium Chloride', *J. Phys. Chem.*, **100**, 1144, 1996.

K. Kajiwara, F. Franks, P. Echlin and A.L. Greer, 'Structural and Dynamic Properties of Crystalline and Amorphous Phases in Raffinose–Water Mixtures', *Pharm. Res.*, **16**, 1441, 1999.

L. Streefland, T. Auffret and F. Franks, 'Bond Cleavage Reactions in Solid Aqueous Carbohydrate Solutions', *Pharm. Res.*, **15**, 843, 1998.

F. Franks, 'Phase Changes and Chemical Reactions in Solid Aqueous Solutions: Science and Technology', *Pure Appl. Chem.*, **69**, 915, 1997.

B.J. Aldous, A.D. Auffret and F. Franks, 'The Crystallisation of Hydrates from Amorphous Carbohydrates', *Cryo-Letters*, **16**, 181, 1995.

S.L. Shamblin, X. Tang, L. Chang, B.C. Hancock and M.J. Pikal, 'Characterization of the Time Scales of Molecular Motion in Pharmaceutically Important Glasses', *J. Phys. Chem.*, **103**, 4113, 1999.

N. Guo, I. Puhlev, D.R. Brown, J. Mansbridge and F. Levine, 'Trehalose expression confers desiccation tolerance on human cells', *Nature Biotechnol.*, **18**, 168–171, 2000.

Chapter 14

W. Lorch, ed., *Handbook of Water Purification*, McGraw-Hill, London, 1981.

L.W. Weisbecker, *The Impacts of Snow Enhancement*, University of Oklahoma Press, Norman, OK, 1967.

Chapter 15

The Independent on Sunday, 12 December 1999.

The Observer, 6 December 1998.

The Independent, 23 October 1999.

The New York Times, January 9 1997.

The Sunday Times, 31 August 1997.

Subject Index